A series of student texts in

CONTEMPORARY BIOLOGY

General Editors :

Professor E. J. W. Barrington, F.R.S.
Professor Arthur J. Willis
Professor Michael A. Sleigh

An Introduction to
Systems Analysis:
with ecological applications

John N. R. Jeffers

F.I.S., F.I.Biol., M.B.I.M.

Director of the Institute of Terrestrial Ecology,
Natural Environmental Research Council, Grange-over-Sands, Cumbria

University Park Press

© John N. R. Jeffers

First published 1978
by Edward Arnold (Publishers) Limited
41 Bedford Square, London WC1B 3DP

First published in the USA by
University Park Press
233 East Redwood Street
Baltimore,
Maryland 21202

Library of Congress Cataloging in Publication Data

Jeffers, John Norman Richard.
An introduction to systems analysis, with ecological applications.
  (A series of student texts in contemporary biology)
  Bibliography: p.
  Includes index.
    1. Ecology—Mathematical models.   2. Systems analysis.
I. Title.
QH541.15.M3J43   574.5'01'84   78-16920
ISBN 0-8391-1305-6

Printed in Great Britain

Preface

The growing interest in systems analysis in ecology is perhaps matched only by the lack of knowledge of what is meant by the term 'systems analysis'. Indeed, many ecologists feel strongly opposed to what they believe to be a concept of systems analysis as applied to problems of research and management of ecological systems. To make matters worse, because this new branch of ecology is relatively young, there is no definitive text book to which the research worker or student can refer. While several texts exist which purport to describe the results of systems analysis as applied to particular problems in ecology, there is little agreement between these texts as to the meaning and scope of the term 'systems analysis'.

This book does not attempt to provide the definitive text referred to above. It is intended as a practical introduction to systems analysis in the broad field of ecology. As such, it is hoped that it will be useful to students of ecology as an undergraduate text, and, possibly, to those graduate students who have little training or experience in mathematical techniques, but who are turning for the first time to the use of mathematical systems as a practical tool in their research or management. The book may also appeal to biologists and others who have relatively little knowledge of mathematics, but who wish to obtain an insight into the theory and practice of systems analysis.

In consequence, the book makes only a limited use of mathematics, and certainly requires no advanced level of mathematics or statistics for its understanding. Some limited use is made of mathematical formulae, and a little experience with simple statistical calculations will be helpful to the reader. Even more helpful will be knowledge of computer programming in one or more of the higher level languages, for example FORTRAN, ALGOL or BASIC. If, however, any reader is deterred by the limited amount of mathematics included in the book, the introduction may be regarded as having failed, at least for that reader.

At the end of each chapter of this book will be found patterned notes summarizing the contents of the chapter. These follow the ideas presented by Tony Buzan in *Use Your Head*, published by the British Broadcasting Corporation in 1974.[9] The notes may be used as a basis for more detailed notes by readers, as an index to the structure and contents of the chapters, and as an aid to revision and learning.

Grange-over-Sands   J.N.R.J.
1978

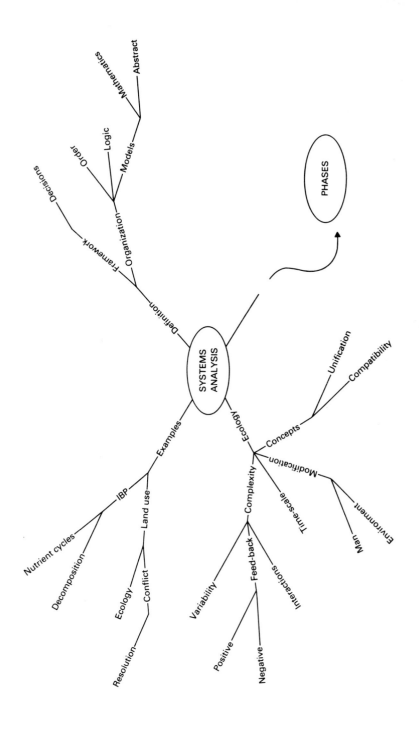

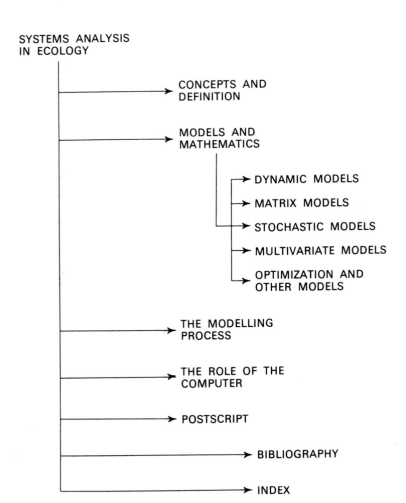

SYSTEMS ANALYSIS
IN ECOLOGY

CONCEPTS AND
DEFINITION

MODELS AND
MATHEMATICS

DYNAMIC MODELS

MATRIX MODELS

STOCHASTIC MODELS

MULTIVARIATE MODELS

OPTIMIZATION AND
OTHER MODELS

THE MODELLING
PROCESS

THE ROLE OF THE
COMPUTER

POSTSCRIPT

BIBLIOGRAPHY

INDEX

Contents

PREFACE v

1. WHAT IS SYSTEMS ANALYSIS? 1

2. MODELS AND MATHEMATICS 12
Word models 12
Mathematical models 14
Deterministic models 15
Stochastic models 16
Practical definitions 17
Simple examples 19
Families of mathematical models 21
Advantages and disadvantages of mathematical models 21

3. DYNAMIC MODELS 24

4. MATRIX MODELS 48

5. STOCHASTIC MODELS 67
Spatial patterns of organisms 67
Analysis of variance 77
Multiple regression analysis 86
Markov models 89

6. MULTIVARIATE MODELS 99
Descriptive models 101
Predictive models 122

7. OPTIMIZATION AND OTHER MODELS   140
   Optimization models   140
   Game theory models   146
   Catastrophe theory models   149

8. THE MODELLING PROCESS   157
   Definition and bounding   157
   Complexity and models   159
   Impacts   160
   Word models   161
   Generation of solutions   162
   Hypotheses   162
   Model Construction   163
   Verification and validation   164
   Sensitivity analysis   166
   Planning and integration   167

9. THE ROLE OF THE COMPUTER   171

   POSTSCRIPT   181

   REFERENCES   183

   INDEX   188

I

What is Systems Analysis?

Contrary to the belief of many ecologists, systems analysis is not a
mathematical technique, nor even a group of mathematical techniques.
It is a broad research strategy that certainly involves the use of
mathematical techniques and concepts, but in a systematic, scientific
approach to the solution of complex problems. As such, it provides a
framework of thought designed to help decision-makers to choose a
desirable course of action, or to predict the outcome of one or more
courses of action that seem desirable to those who have to make
decisions. In particularly favourable cases the course of action that is
indicated by the systems analysis will be the 'best' choice in some
specified or defined way.

In the sense in which we shall use the term in this book, systems
analysis is the orderly and logical organization of data and information
into models, followed by the rigorous testing and exploration of these
models necessary for their validation and improvement. Further
definition of the term 'model' will be given in Chapter 2, but, for the
present, we may regard these models as formal expressions of the
essential elements of a problem in either physical or mathematical
terms. In past scientific work much of the emphasis in scientific
explanation has been on the use of physical analogues of biological and
environmental processes, and, even in systems analysis, we may
occasionally have reference to physical analogues of this kind. More
generally, however, the models of systems analysis will be mathematical and essentially abstract.

Initially, we may identify seven steps in the application of systems
analysis to a practical problem of ecology. These steps and their inter-

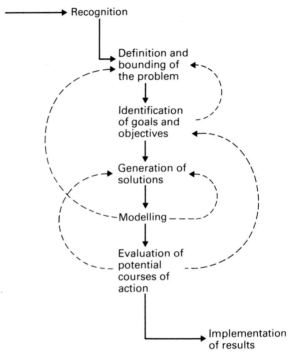

Fig. 1.1   Diagram of the phases of systems analysis.

connection are summarized in Fig. 1.1, and are described briefly
below. They will be considered in greater detail in Chapter 7, when we
discuss the relationship between the stages of systems analysis and the
solution of practical problems. Nevertheless, it is desirable to obtain a
perspective of the various stages before reviewing some of the types of
models which will form an essential core of the systems analysis.

(i) Recognition

The recognition of the existence of a problem, or of a constellation
of interconnected problems, which may be amenable to analysis, and
which is sufficiently important to expose to detailed investigation, is
not necessarily a trivial step. As we know from bitter experience, it
is desperately easy to overlook some practical aspect of ecology which
should be investigated, or to assume that commonly-held beliefs about
ecological processes and systems are true because they are widely held.
However, the recognition that research is necessary is as important
as the choice of the correct method to use in the research. It may be

relatively easy to choose problems for investigation which are not readily amenable to systems analysis. Similarly, it is relatively easy to choose problems which do not require the power of systems analysis for their solution, and for which it would be uneconomic to carry out research by the methods of systems analysis. This double phase of recognition, therefore, may be critical in determining the success or failure of the investigation.

(ii) Definition and bounding of the extent of the problem

Once the existence of the problem has been recognized, it is necessary to simplify it to the point at which it is likely to be capable of analytical solution, while, on the other hand, preserving all the elements which make the problem of sufficient interest for practical research. Again, this is a critical stage in any systems analysis. The difficult judgement of the relative importance of the inclusion or exclusion of elements of the problem, and the balancing of their relevance to the analytical grasp of the situation against their contribution to complications which may well become unmanageable will frequently depend upon experience in the application of systems analysis. This is one stage at which the experienced systems analyst can make his most valuable contribution. The delicate balance between simplification and complexity, whilst retaining sufficient relationship to the original problem for the analytical solution to be recognizably appropriate, will almost certainly determine the success or failure of the investigation. Many promising projects have ultimately proved worthless because the complexity of the problem was allowed to swamp the subsequent modelling, with the result that it became impossible to derive a solution. In contrast, much of the systems analysis that has been carried out in the past in the broad field of ecology has resulted in trivial solutions to a problem which was essentially only a subset of the original problem.

(iii) Identification of the hierarchy of goals and objectives

Once the extent of the problem has been defined and bounded it should become possible to define the goals and objectives of the investigation. Usually, these goals and objectives will form a hierarchy, with the major objectives progressively sub-divided to a series of minor objectives. In such a hierarchy, it will also be necessary to determine priorities for the various stages and to determine priorities relative to the amount of effort that will be required to meet the objectives. Thus, in a complex investigation, the systems analyst may decide to place relatively little priority on goals and objectives which, while desirable from the point of view of scientific information, have little effect upon the kind of decisions that may need to be made about the management or treatment of the ecological system. Alternatively, where the investigation forms part of a programme of fundamental science, he may be prepared to accept certain defined alternative forms of management and concentrate most of the effort on objectives relevant to the ecological processes themselves. For a successful systems analysis, however, it is important that the priorities that have been assigned to various objectives should be defined.

(iv) Generation of solutions

At this point, it will usually become possible for the investigator to generate a series of possible solutions to the problem. Further discussion of the way in which these alternatives arise will be given in Chapter 2, but, broadly, an experienced systems analyst will recognize families of possible solutions to specific problems. In general, he will seek an analytical solution of the greatest possible generality because, in this way, he can make the best use of previous work on similar problems and the underlying mathematics of their solution. It is only rarely that any specific problem has only one possible method of solution. Again, the experience of the analyst will be helpful in choosing the most appropriate family from which to derive his analytical solution. An inexperienced systems analyst may waste a considerable amount of time and money seeking to apply a solution from an unpromising family, without recognizing that the solution that he has chosen makes assumptions which are unjustified in the particular case with which he is dealing. Frequently, the analyst will deliberately develop several alternative solutions before deciding on the one which is most appropriate to his problem.

(v) Modelling

When the appropriate alternatives have been examined, the important phase of modelling the complex, dynamic interrelationships between various facets of the problem can begin. Such modelling must be done with a full awareness of the inherent uncertainties in the various processes to be modelled, and of the feed-back mechanisms which may considerably complicate both the understanding and the tractability of the system. Similarly, the modelling must itself bear in mind the complex series of rules which will necessarily be used in reaching a decision about appropriate courses of action. It is very easy, at this stage, for the mathematician to be carried away by the ingenuity and elegance of his model, with the result that all contact is lost between the reality of the decision-making process and the mathematics used to determine the possible consequences of the decisions.

(vi) Evaluation of potential courses of action

Once the modelling has been brought to a sufficiently advanced stage for the model to be used, at least in a preliminary way, the phase of evaluating potential courses of action from the model can begin. In the course of this evaluation it will be necessary to investigate the sensitivity of the results of the assumptions made by the model, as it is only when the model begins to be used that previously unsuspected weaknesses in the assumptions and in the model formulation begin to be revealed. Discovery of an important flaw in a major assumption may lead to a return to the modelling phase, but, frequently, further progress can be made by relatively simple modifications to the original model. It is also usually necessary to investigate the sensitivity of the model to facets of the problem which were excluded from the formal analysis in the second phase, when the problem and its extent were defined and bounded.

(vii) Implementation of the results

The final phase in the systems analysis is the implementation of the results which have been derived from the previous phases. If the analysis has been carried through in the way described above, the steps necessary for the implementation of the results will usually be fairly obvious. Nevertheless, the systems analysis is not complete unless the analysis has moved to the implementation phase, and it is in this respect that many of the past attempts at systems analysis have been incomplete. The implementation may itself demonstrate that various phases of the analysis were incomplete or need to be revised, so that a degree of recycling through some of the phases already completed may be necessary.

Because systems analysis is a framework of thought rather than a defined prescription, the list of phases above needs to be understood in a qualified sense. Not all of the steps need to be included in every instance of a systems analysis, and it may be possible to exclude some phases in any particular case. Similarly, the order in which the phases may be undertaken may be varied or it may be necessary to iterate through them in various patterns. For example, the importance of excluded factors may have to be reassessed repeatedly, necessitating several cycles of the modelling and evaluating phases. Similarly, the relevance of the objective structure of the analysis may have to be examined periodically, sometimes requiring a return to one of the early phases even after a considerable amount of work has been done on some of the middle and later phases. The most useful models will mimic reality with sufficient precision to serve a broad spectrum of decisions and decision-makers. The decision phase may, therefore, be diffuse and broad, and follow the completion of the formal scientific analysis.

The aim of the broad framework of systems analysis outlined above is to promote good decision-making in practical applications, and, in our case, in ecology. The framework is intended to focus and to force hard thinking about complex, and usually large, problems not amenable to solution by simpler methods of investigation, for example by direct experimentation or by survey. Because of the complexity of the problems to which systems analysis is usually applied, the analysis frequently involves the use of electronic computers for data processing and analysis, and of complex mathematics for the resolution of choices between alternative solutions; but the use of either computers or mathematics, or both, is not an essential feature of systems analysis itself. Indeed, the framework can sometimes best be illustrated by problems which do not require the use of either.

The special contribution of systems analysis to problem-solving lies in the identification of unanticipated factors and interactions that may subsequently prove to be important, in the forcing of modifications to experimental and survey procedures to include these factors and interactions, and in illuminating critical weaknesses in hypotheses and assumptions. Just as the scientific method, with its insistence on the test of hypotheses through practical experiments and rigorous sampling procedures, provides the essential tools for advances in our knowledge of the physical world, systems analysis welds these tools into a flexible but rigorous exploration of complex phenomena. Success in its application to practical problems is most likely to come from small groups of scientists working within a single institute and with a well-defined and rather narrow focus. The conditions necessary for such groups are discussed in some detail in Chapter 7, and it is sufficient to stress here that a successful systems analysis group will, from the start, have made a particular effort to link the modelling phase with a carefully designed research strategy and with vigorous validation of all the data used in the analysis.

Having defined systems analysis in general terms, why should we need to use systems analysis in ecology? In part, the answer to this question lies in the relative complexity of ecology as a science, dealing as it does with the many-sided interactions of a wide variety of organisms. Nearly all of these interactions are dynamic, in the sense that they are time-dependent and constantly changing. Furthermore, the interactions frequently have the feature which the engineer calls 'feed-back', i.e. the carrying back of some of the effects of a process to its source or to a preceding stage so as to strengthen or modify it. Such feed-back will sometimes be positive, in the sense that the effect is increased, and sometimes negative, in the sense that the effect is decreased. The feed-back may itself be complex, involving a series of positive and negative effects, with various results depending upon a series of environmental factors.

The complexity of ecosystems is not, however, confined to the presence of multiple interactions in the relationships between organisms. Living organisms are themselves variable—indeed, variability is one of their essential characteristics. This variability may be expressed in terms of effects on other organisms, for example by competition or by predation, or it may be expressed in the response of the organisms, either collectively or singly, to environmental conditions. Such response will be reflected in variable rates of growth, and of reproduction, or even in variable ability to exist under markedly adverse conditions. When this characteristic is added to independent variations in environmental factors such as climate and habitat, ecological processes and ecological systems become difficult to investigate and control.

As a result, the understanding of even relatively unmodified ecological systems is far from easy. The traditional response of the ecologist to the difficulty has been to focus attention on to small subsets of the real problem. Much research has been concentrated on the behaviour of single organisms in simplified habitats, for example on flour-beetles in bags of flour or enchytraeid worms on selected media. Alternatively, the competition between two or three species, again in relatively simple habitats, has been studied extensively.

An especially popular type of ecological research is on the predator-prey relationships between one predator and one prey, as for example between the deer-mouse, *Peromyscus leucopus*, and sawfly larvae in the laboratory,[36] or *Paramecium* and *Didinium*.[53] In all of these examples an attempt has been made to reduce the level of the complexity studied to a level which is manageable by traditional methods of investigation, by eliminating many of the possible sources of variability. Even when this has been done, the interrelationships remain difficult to model and to understand.[55]

When the effects of deliberate modification of ecological systems are included in ecological research, a further dimension of variability and interaction is introduced. In the important topics of applied ecology of forestry and agriculture, some simplification of the ecological system is usually achieved by considering the response of the crop species alone, but such research provides very little information on the response of the system as a whole to modifications introduced by changes in management. In particular, the effects of the crop species on the soil, and on species of organisms associated with the ecosystem on which the crop has been imposed, are seldom studied, mainly because of the difficulty of designing experiments which are capable of testing hypotheses with the necessary degree of complexity. The extension of these ideas to the ecological effects of land use, where several alternative strategies for land use and environmental management are considered, is even more difficult, and therefore seldom attempted. Research on the deliberate management of natural or semi-natural ecosystems, for example on the management of nature reserves to ensure the conservation of wildlife, has also seldom been attempted, again because of the difficulty of encompassing the complexity and variability of the many species contributing to the stability or instability of the ecosystem.

For all the above reasons, i.e. the inherent complexity of ecological relationships, the characteristic variability of living organisms, and the apparently unpredictable effects of deliberate modification of ecosystems by man, the ecologist requires an orderly and logical organization of his research which goes beyond the sequential application of tests of hypotheses, although the 'appeal to nature' invoked by the experimental method necessarily remains at the heart of the organization. Applied systems analysis provides one possible format for that organization, a format in which the experimentation is embedded in a conscious attempt to model the system so that the complexity and the variability are retained in a form in which they are amenable to analysis. Exponents of systems analysis do not claim that their approach to the solution of complex problems is the only approach that is possible, but, understandably, they expect it to be the most effective approach— if a more effective solution to the problem existed, they would use it!

There is, however, a further reason for the use of systems analysis in ecological research. By its very nature, ecological research frequently requires long time-scales. Agricultural and horticultural research, for example, is largely concerned with crops which are harvested annually, so that one cycle of experimentation takes a year or more to complete. The search for an optimum level of fertilizer application, possibly combined with other cultural operations, may therefore take several years, especially when the interaction between the experimental treatments and the weather need to be considered. In forestry, because of the longer rotations of tree crops, a short-term experiment may easily last 25 years, and long-term experiments may last for anything from 40 to 120 years. Research on resource management frequently involves similar time-scales, and experimental procedures are relatively slow. It is, therefore, necessary to ensure the greatest possible advance from each stage of the experimentation, and the models of systems analysis provide the necessary framework.

Then, again, the present state of ecology as a science, with its extremely scattered research effort over a wide field, urgently needs a unifying concept. Not only is there a marked incompatibility of the many existing theories, but the weakness of the assumptions behind these theories is largely unexplored, partly because the assumptions themselves have never been stated. There are many other branches that are in a similar phase, but ecology is certainly one of the branches in which systems analysis can act as a filter, though not the only filter, of existing ideas. Theories which have been shown to be incompatible can be tested as alternative hypotheses, and systems analysis will itself frequently suggest the critical experiments necessary to discriminate between these hypotheses.

Finally, we need to consider carefully the nature of the models which we will wish to construct of ecological relationships. It is not usually recognized how much of our thinking of relationships in the physical sciences is conditioned by the functional models of engineering and physics. Indeed, as we shall see in the next chapter, many of the families of models which we may seek to employ in systems analysis will be of the functional and deterministic kind derived from the causal relationships of physics, and the mathematics used to describe these models is the traditional applied mathematics we all learn in school, and which is, strictly speaking, mathematics applied to physics. Many ecological relationships are, however, not of this kind, involving, as we have seen, the variability of organisms and of habitats, as well as the interaction between organisms and habitats. The methods of direct investigation and experimentation become difficult to apply, and the more complex mathematics of stochastic or probabilistic relationships need to be invoked to model the variability of response to biological processes and the directional relationships of variables dependent on a series of apparently independent variables.

So far, relatively little of the application of systems analysis to ecology has reached the stage of publication, so that practical examples are not easy to quote. Two examples may, however, be cited as illustrations of the ways in which systems analysis may be used in ecology. First, we will examine briefly the ways in which systems analysis has been used in the investigation of nutrient and decomposer cycles in the research of the International Biological Programme, and, second, we will look at the role of systems analysis in studies of the ecological effects of land use.

Research on nutrient and decomposer cycles of ecological systems is far from easy. Direct experiments on the processes involved in the cycles are difficult because of the complexity of the cycles—simultaneous determination of the many parameters involved may even be impossible without disturbing the processes themselves, especially over the long time-scales necessary for the detection of seasonal and periodic variations. Determination of the parameters of parts of these processes in successive phases may be unsatisfactory because of seasonal or annual variations from phase to phase, and because of the lack of any effective measurement of interactions between the parts in successive phases. In the IBP programmes, considerable progress was nevertheless made in the synthesis of models of nutrients and decomposer cycles for particular biomes through the use of systems analysis. Many of the models were derived initially from data accumulated from past research, leading to critical tests suggested by direct simulation of the processes, and thence to further data collection and modelling. In Britain, for example, significant advances were achieved in the modelling of semi-natural deciduous woodland and of upland moorland, and, for the latter, a preliminary synthesis of tundra systems has taken international research to the stage at which studies of the effects of management or exploitation of tundra have become feasible.

Similarly, research on the effects of major changes in land use on terrestrial ecosystems has seldom been attempted because of the large areas of land concerned, and the long time-scales of the research. Again, the synthesis of ecological models, beginning from past research results, through critical experiments, has enabled predictions to be made of the effects of changes in management, and hence of the effects of defined land use policies. One of the interesting features of the early attempts at this kind of synthesis is the recognition that very detailed models of the ecology of the component ecosystems are not the first priority. The most urgent need is for relatively simple models which are capable of recognizing the potential conflict between the land use policies of the various agencies having an impact on the area considered. Frequently, it is possible to formulate the models so that a 'game-playing' situation can be set up between the various agencies to enable them to recognize the conflict between their policies, and to explore ways of resolving the conflict—a kind of 'Monopoly' for land use.

The ways in which systems analysis can be used in examples such as these will be explored in greater detail in Chapter 7 when we return to the strategic considerations of the modelling process. Before turning to this aspect of systems analysis, however, we will need to explore the role of models as an essential element of the framework and review some of the commoner families of models.

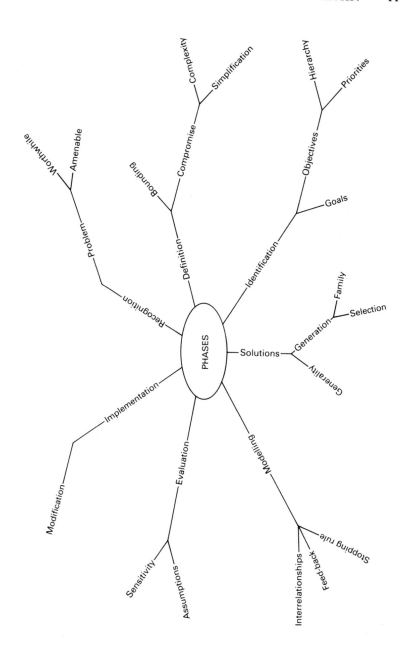

2

Models and Mathematics

In the last chapter, we described models as formal expressions of the essential elements of a problem in either physical or mathematical terms. In this chapter, we will need to give further definition to this description so as to obtain an understanding of what is involved in the important modelling phase of systems analysis and in the choice of an appropriate model for the solution of the problem. We will begin by an examination of what is implied by the term 'formal expressions' and by the restriction of these expressions to physical or mathematical terms.

WORD MODELS

The non-mathematically orientated reader may well ask 'Why do the essential elements of a problem have to be given a formal expression, and why, in particular, do we restrict this expression to physical or mathematical terms? Can we not proceed equally well from a purely verbal description of the problem, and avoid this preoccupation with mathematics?' It is true, of course, that our first recognition of the problem is likely to be expressed in verbal terms, and there is much to be gained by seeking the most precise verbal description that we can find of the problem with which we are concerned. Indeed, some systems analysts explicitly recognize the 'word model' as an essential preliminary to the modelling phase, in which all concerned with the problem to be solved combine to produce an agreed description of the ecological system in words, with particular emphasis on that part of the system which is thought to be relevant to the investigation. It is surprising how often even four or five people closely concerned with a problem will disagree with each other's descriptions of the same ecological system, and disagreement on the particular elements of the systems which contribute, directly or indirectly, to the problem of practical concern is even more likely. For the larger groups likely to be encountered in the complex problems for which systems analysis is appropriate, the disagreement may be both striking and difficult to resolve. There is, therefore, all the more reason to spend some time in an attempt to find an agreed description, even if that description contains some passages, expressed as alternatives, for which no agreement can be reached. Such a description may well help in the phases of definition and bounding of the extent of the problem and of identification of the hierarchy of goals and objectives. To this extent, a 'word model', although not included in the definition of 'model' used in this book, may well be invaluable.

Mellanby[57] objects to the term 'word model'. He points out, quite correctly, that all that is implied is a description and that there is therefore no need to invest the perfectly ordinary process of describing something with a complicated name. However, as the definition of model used in this text implies a formal expression, it is useful to be able to distinguish between these formal expressions and purely verbal descriptions of events, processes and relationships.

It is only fair to point out, however, that many experienced exponents of systems analysis do not regard emphasis on 'word models' as worthwhile, except in so far as such models concentrate on the definition and bounding of the problem and on the identification of the hierarchy of goals and objectives. In part, this is because the experienced analyst may well be able to proceed quickly to the generation of a series of possible models, and will regard the selection of the most appropriate of these models as more likely to lead to a feasible solution. In part, disenchantment with 'word models' comes from the inherent difficulties of the models themselves, and it is these difficulties that we must now examine.

In essence, there are two major difficulties, all others stemming directly from them. First, word models quickly run into problems when they seek to define complex relationships. It is true that even the most subtle relationships *can* frequently be expressed by words, and we have the whole wealth of literature in many languages as evidence of this, but the simultaneous operation of complex interrelationships in ecological systems is not easily represented by sequential relationships between words. The essential features of the relationships quickly become lost in the many words required to describe them, and nowhere more quickly than in the description of feed-back, which we have already defined as the carrying back of some of the effects of a process to its source or to a preceding stage so that the effect itself is increased or decreased. If any ecologist doubts this statement of the difficulty of using word models to describe biological processes, he may care to try to describe in words the interrelationships between photosynthesis and respiration in plants, as expressed by the exchange of oxygen and carbon dioxide.

Second, it is unfortunate that the same word will not necessarily have the same meaning for each member of the group concerned with the problem. Words such as 'biomass', 'increment', and 'standing crop' all need careful definition before they can be incorporated into the model. Having been defined, the word then becomes a symbol, but if anyone outside the original group begins to use the model, he may well interpret the word in a wider context than the narrower definition implied by the symbol.

MATHEMATICAL MODELS

The strength of mathematics lies in its ability to provide a symbolic logic which is capable of expressing ideas, and particularly relationships, of very great complexity, while, at the same time, retaining a simplicity and parsimony of expression. Admittedly, to the non-mathematician, 'simplicity of expression' may seem a most unlikely property of mathematics, but the whole basis of mathematical notation rests on the economical expression of relationships as a symbolic logic, and this expression is 'formal' in the sense that it enables predictive statements to be derived from the relationships. Without the ability to predict the result of changes in one or more of the elements in the relationship, we could not regard the expressions as science rather than metaphysics or literature.

Our use of mathematical notation in the modelling of complex systems is therefore an attempt to provide a representational symbolic logic which simplifies, but does not markedly distort, the underlying relationships. Logicians sometimes speak of this use of symbolic logic as the 'mapping' of a system by the use of a homomorph, i.e. an imperfect representation of reality, a caricature of reality.

The various mathematical rules for manipulating the relationships enable us to derive predictions of the changes which we may expect to occur in the ecological systems as various component values of these systems are changed. These predictions, in turn, enable us to make comparisons between our model systems and the real systems which they are intended to represent, and, in this way, to test the adequacy of the model against observations and data derived from the real world—the 'appeal to nature' which is necessary for the application of the scientific method. Indeed, manipulation of the model system may itself suggest the experiments which are necessary to test the adequacy of the system.

Maynard Smith[56] makes a distinction between 'models' and 'simulations'. He regards a mathematical description with a practical purpose, and which includes as much relevant detail as possible, as a 'simulation', and restricts the use of the term 'model' to descriptions of general ideas which include as little detail as possible. This is not a distinction which will be made in this text. Any formal expression of the relationships between defined symbols will be regarded as a model, and such models will then generally be used to simulate the behaviour of the ecological system resulting from changes induced in the system. In the applied systems analysis, therefore, we will be attempting to achieve the fusion of a 'model' and a 'simulation' in Maynard Smith's terms.

DETERMINISTIC MODELS

Before we can proceed very much further with this general consideration of models, we will need some working definitions to cover essential concepts. It will also be helpful, however, to have some examples of simple models on which to hang these definitions, and some will therefore now be presented. One of the simplest models of growth of a population of organisms is that given by the differential equation:

$$\frac{dy}{dt} = ry$$

where y is the density of the population at time t, and r is a constant. The growth of a bacterial colony before the medium is exhausted is one example of a biological process which may be represented by such a model, in which the rate of growth at any point in time is a constant proportion of the density of the population at that time. Merely by expressing the relationship in this form, we can use the properties of the particular kind of symbolic logic represented by differential equations to show that the density of the population at any point in time may also be represented by the equation:

$$y = y_0 e^{rt}$$

where, again, y is the density of the population at time t, y_0 is the density at time $t = 0$, r is a constant, and e is the base of the natural logarithms.*

This simple exponential model has a rather limited usefulness, as an increasing population of organisms will usually exhaust its resources, and so settle down to some steady density. An alternative model, possessing this property, is the differential equation:

$$\frac{dy}{dt} = ay - by^2$$

where y is again the density of the population at time t, and both a and b are constants. Again, this model may also be represented by the equation:

$$y = \frac{a/b}{1 + e^{-a(t - t_0)}}$$

where y is the population density at time t, y_0 is the population density at time $t = 0$, a and b are constants, and e is the base of the natural logarithms. This logistic model represents quite well the growth of bacterial populations with a limited supply of resources. The population at first follows an exponential pattern of growth which is gradually reduced as the resources become more limited until the size of the population reaches a constant level, or asymptote. Furthermore, we can predict that the constant level will be a/b by simple algebraic manipulation of the original equation of the model, as a logical deduction within the formal symbolic logic of the mathematical expression of the model. In other words, expressing the model in abstract mathematical terms, instead of words, has immediately enabled us to derive further information from the model.

STOCHASTIC MODELS

Both models mentioned above are 'deterministic' in the sense that, given the values of the constants, the density of the population at a given time t is always the same—the value of y is completely determined by the value of t, in the familiar physical sense of the cause always being followed by the same effect. Differential equation models were indeed originally developed in the application of mathematics to

---

* Any reader not familiar with the algebra of differential equations will have to take this statement on trust. The main point is that there is a well-established logical dependence between the two equations and writing the model in one of these forms automatically implies the other.

physics—the traditional meaning of applied mathematics—and it is only natural that, in the search for models in ecology, we should first explore the possible application of what has already been developed in other fields.

We may, however, formulate our models in a very different way, exploiting the variability of living organisms as a basic property of the model, in probabilistic or stochastic models. Such models employ a quite different branch of mathematics, more recent in its development than that of differential equations and the calculus. A simple example of such a model, corresponding to the deterministic model of exponential growth, is:

$$\frac{dy}{dt} = [a + y(t)]y$$

where y is the density of the population at time t, a is a constant, and $y(t)$ is a random variable with mean zero. This means that the value of $y(t)$ is different from occasion to occasion, being taken from a random distribution so that there is no correlation between the fluctuations at successive instants. The assumption of no serial correlation may seem unrealistic, but all it means is that the fluctuations are correlated only over times which are short compared to other time-scales relevant to the system.

It is easy to see that, if a stochastic model is used as a basis for a simulation, the outcome of the simulation will not always be the same, even when the constants and the starting values are the same. The random elements in the model will provide variability, and the aim of such models is to mirror the variability found in living organisms and in ecological systems. As with experiments on the organisms themselves, it will usually be necessary to undertake repeated trials of the simulations in order to determine the ways in which the system will respond to various changes.

PRACTICAL DEFINITIONS

In our preliminary examples, therefore, we have briefly examined two kinds of models, or sets of rules for computing predicted values, with which observed values can be compared:

1. Deterministic models, for which the predicted values may be computed exactly.
2. Stochastic models, for which the predicted values depend on probability distributions.

It is important that this distinction is kept in mind, because we will
next have to examine the way in which probability distributions are
used in the fitting of models, that is in the selection of the values of
constants that generate predicted values which are acceptably similar
to the observed values. Unless we have some prior knowledge of the
constants in our equations, they will need to be estimated from samples
drawn from experiments and surveys. The estimation of such constants requires the use of statistical techniques which themselves
depend upon probability theory, and we will need some practical
definitions to aid our understanding of models.

First, we will continually need to distinguish between a ***population***
and a ***sample*** drawn from that population. We define the total set of
individuals about which inferences are to be made as a ***population.***
These individuals may be organisms, ecosystems, quadrats, or indeed
any measure or characteristic of organisms or ecosystems. A ***sample***
is any finite set of individuals drawn from that population, and we will
assume that the sample is taken in such a way that values computed
from the sample are representative of the complete population and
may be regarded, therefore, as estimates of the values for the population. In this text, we will not describe the methods of sampling which
are necessary to obtain unbiased estimates of population values—any
good statistical text-book will provide the necessary methods—but we
will assume that appropriate methods of sampling are used to derive
samples which are representative of some defined population. Characteristic values of populations will be defined as ***parameters***, in
contrast to the corresponding values of samples which will be regarded
as estimates of those parameters, either as ***constants*** or ***coefficients***
in model equations. We will always need to be aware of the difference
between parameters and sample statistics.

Second, our model equations will contain two kinds of variables.
One, at least, of the variables will be dependent, in the sense that it is a
variable which is expected to be altered by changes in other variables.
In our examples, the density, y, is a dependent variable. The other
variables are regarded as ***independent*** variables in the language of
the differential calculus, in the sense that it is the alteration of these
variables which leads to the change in the dependent variable. This
use of the word '***independent***' is, however, misleading as two or more
independent variables may indeed be strongly correlated. Whenever
we use the techniques of regression analysis to provide estimates of
model parameters, we will use the term '***regressor***' variables to
describe those variables which provide the changes necessary to
induce changes in the dependent variable.

Third, we will define ***model fitting*** as the selection of parameters that generate predicted values acceptably similar to the observed values. Indeed, we will regard the probability that the parameters give rise to the observed data as a mathematical function of the parameters, defined as the **likelihood function**. The function is a measure of agreement between model and data, and the parameter values for which the likelihood is a maximum are known as **maximum likeli-hood estimates**. In practice, our models will fall into two further categories, i.e. **analytic models** and **simulation models**. **Analytic models** are those for which explicit formulae are derived for predicted values or distributions—they include regression and multivariate models, experimental designs, and the standard, theoretical statistical distributions. **Simulation models** are those that can be specified by a routine of arithmetic operations, such as the solution of differential equations, the repeated application of a transition matrix, or the use of random or pseudo-random numbers. Simulation models have the advantage of being easier for the non-mathematician to construct, but they are usually much more difficult to fit to observed data than analytic models.

SIMPLE EXAMPLES

In the light of these definitions, let us now look at some more examples of relatively simple mathematical models in ecology. Volterra[88] described the interactions between a prey species, with a density x, and its predator, with a density y, by the differential equations:

$$\frac{dx}{dt} = ax - bx^2 - cxy$$

$$\frac{dy}{dt} = ey + c'xy$$

where a, b, c, e, and c' are coefficients
x is the density of the prey species
y is the density of the predator species
t is time.

Clearly, this is a deterministic model, in that, for given starting values of x and y, and for given values of the coefficients, we will always arrive at the same predictions. The dependent variables are x and y, and the independent variable is t (note that the format of differential equations frequently obscures the identification of the nature of variables) but the prediction of x and y also depends on previous values of x and y so that the model is said to be 'recursive'.

The behaviour of the model is relatively well-known, and is amenable to mathematical analysis, so that this is an analytic model rather than a simulation model. In the absence of predation, the prey species would follow the logistic equation, with an intrinsic rate of increase, a, and a carrying capacity, a/b. The rate at which prey are eaten is proportional to the product of the densities of both predator and prey. It is also possible to show, by mathematical analysis alone (see, for example, Maynard Smith[56]) that, if the carrying capacity of the prey is high enough to support the predator, both predator and prey numbers oscillate with decreasing amplitude, the predator oscillations lagging in phase behind the prey.

Of course, for any defined population of predator and prey, we will not know the exact values of the population parameters α, β, γ, η and γ' (note the useful convention of signifying population parameters by Greek letters, in contrast to the sample estimates in lower case letters) and we will need to find some way in which the estimates of the population parameters can be derived from sample data. An example of the fitting of a model of this kind will be given later.

An interesting example of a stochastic model was given by Skellam[72] as a possible explanation of variation in the number of species occurring in a genus for a wide variety of groups of living organisms. He sets up a hypothesis that the evolutionary tree is the outcome of a stochastic process in which, at intervals, each line of descent is liable to take one of three mutually exclusive courses:

1. To remain unbranched
2. To branch into two
3. To disappear from the genus to which it belongs by
    (a) becoming extinct
    (b) becoming the starting point of a new genus (i.e. changing its generic name).

From this hypothesis, it is possible to derive the equation:

$$\Phi_{n+1}(u) = \Phi_n(\alpha u + \beta u^2)$$

where Φ_n is the factorial moment generating function of the number of species in a genus which originated exactly n years ago. It can be shown that the solution of this equation leads to a geometric distribution with an abnormal zero-th class. Integrating this distribution over n, we arrive at the prediction that the number of species in a genus should follow the logarithmic distribution, and the numbers of species in known genera follow this distribution quite closely. Note that, in this example, we have not even had to estimate the values of the parameters α and β—the argument rests on the deductive logic of the mathematics once the form of the model has been decided. If, however, we had wanted to predict the number of species in particular genera, we would have needed some method of estimating the population parameters.

FAMILIES OF MATHEMATICAL MODELS

Some of the general properties of mathematical models have been touched on here; subsequent chapters in this book will examine particular families of models in more detail. It is one of the added benefits of the use of mathematical models that an experienced analyst can recognize 'families' of models, in much the same way that an experienced botanist is often able to place a plant into a genus even when he does not know the species. It would be impossible, in an introductory text of this kind, to include all the possible families, and our examination will be confined to those which the user of systems analysis is most likely to meet, namely:

1. Dynamic models
2. Compartment models
3. Matrix models
4. Multivariate models
5. Optimization and other models.

As we shall see, this list is far from being exhaustive, and the categories are also not mutually exclusive. The classification is, however, sufficient to provide us with some examples of mathematical models applied to real problems, and to illustrate the basic requirements of models in applied systems analysis.

ADVANTAGES AND DISADVANTAGES OF MATHEMATICAL MODELS

Before turning to the examination of these generic families of mathematical models, however, let us summarize the relative advantages and disadvantages of the use of mathematical models in applied systems analysis.

The advantages of mathematical models are that they are precise and abstract, that they transfer information in a logical way, and that they act as an unambiguous medium of communication. They are precise, because they enable predictions to be made in such a way that these predictions can be checked against reality by experiment or by survey. They are abstract, because the symbolic logic of mathematics extracts those elements, and only those elements, which are important to the deductive logic of the argument, thus eliminating all the extraneous meanings which may be attached to words. Mathematical models transfer information from the whole body of knowledge of the behaviour of interrelationships to the particular problem being investigated, so that logically dependent arguments are derived without the necessity for all the past research to be repeated. Mathematical models provide a valuable means of communication because of the unambiguity of the symbolic logic employed in mathematics, a medium of communication which is largely unaffected by the normal barriers of human language.

The disadvantages of mathematical models lie in the apparent complexity of the symbolic logic, at least to the non-mathematician. In part, this is a necessary complexity—if the problem under investigation is complex, it is likely, but not necessary, that the mathematics needed to describe the problem will also be complex. There is also a certain opaqueness of mathematics, and the difficulty that many people have in translating from mathematical results to real life is not confined to the non-mathematicians. The failure to interpret correctly the results of advanced methods of analysis is evident in many papers submitted to scientific journals, possibly because the interpretation of the results of analysis is far less frequently discussed than the underlying mathematics of the analysis. Bross[8] makes the point that 'pseudo science flourishes because relatively few people take the trouble to translate from algebra into an everyday language where the absurdity of the argument can be clearly seen'.

Perhaps the greatest disadvantages of mathematical models, however, are the distortion that may be introduced into a problem by rigid insistence on a particular model, even where it does not really fit the facts, and the difficulty which is sometimes apparent in abandoning a model which is no longer capable of advancing the research. Mathematical modelling is an intoxicating pursuit, so much so that it is relatively easy for the modeller to abandon the real world and to indulge himself (herself) in the use of mathematical languages for abstract art forms. It is for this reason that applied systems analysis stresses that 'modelling' is only one of the steps in the broad framework of investigation. We must always be careful not to allow the modelling to become the purpose of the exercise!

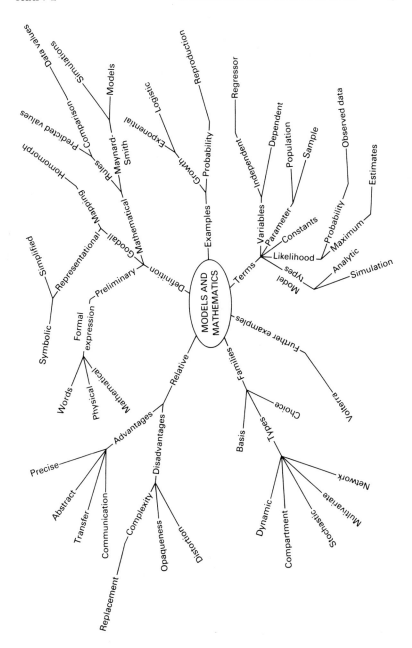

3

Dynamic Models

We will begin our review of some of the families of mathematical models by considering one of the more recent developments in the field of modelling. The study of systems dynamics is based on servo-mechanism theory, itself a relatively recent development, and the use of dynamic models in any practical application depends on the ability of modern high-speed digital computers to solve large numbers (hundreds) of equations in short periods of time. The equations are more or less complex mathematical descriptions of the operation of the system being simulated, and are in the form of expressions for levels of various types which change at rates controlled by decision functions. The level equations represent accumulations within the system of such variables as weight, numbers of organisms and energy, and the rate equations govern the change of the levels with time. The decision functions represent the policies or rules, explicit or implicit, which are assumed to control the operation of the system.

As emphasized in the last chapter, mathematical models of a system can only represent that system to the extent that the equations describing the operation of the components of the model accurately describe the operations of the components of the real system. The popularity of dynamic models arises from the very great flexibility of the methods used to describe system operations, including non-linear responses of components to controlling variables and both positive and negative feed-back. As we will see, however, this flexibility has its disadvantages. It is, in any case, usually impossible to include equations for all the components of a system, as, even with modern computers, the simulation rapidly becomes too complex. It is, there-

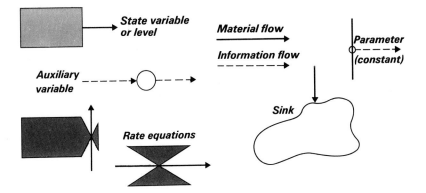

Fig. 3.1   Conventions suggested by Forrester for the presentation of system interrelationships.

fore, necessary to obtain an abstraction based on judgement and on assumptions as to which of the many components are those which control the operation of the system.

The application of systems dynamics in modelling involves three principal steps. First, it is necessary to identify the dynamic behaviour of the system which is of interest, and to formulate hypotheses about the interactions that create the behaviour—we may identify this step with problem definition, and it is clearly closely related to the phases of systems analysis which precede the modelling phase itself, except that we are now concerned with the detailed behaviour and interactions of the variables of the system. Second, a computer simulation model must be created in such a way that it replicates the essential elements of the behaviour and interactions identified as essential to the system. Third, when we are satisfied that the behaviour of the model is sufficiently close to that of the real system, we then use the model to understand the cause of observed changes in the real system and to suggest experiments to be carried out in the evaluation of potential courses of action, i.e. in the succeeding phase of the systems analysis.

One of the attractions of dynamic models for research workers lies in the use of relational diagrams to summarize the main interrelations of a complex system. The conventions introduced by Forrester[24] are given in Fig. 3.1, and, although these conventions were first developed for the presentation of industrial systems, they are equally convenient and applicable to ecology. There are, in any case, considerable advantages to be gained from adopting a standard set of symbols over a wide range of applications.

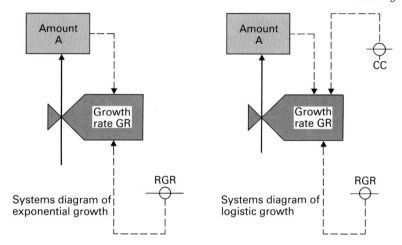

Fig. 3.2 Simple examples of systems diagrams for exponential and logistic growth (see text for explanation).

Forrester assigned special symbols to the various types of variables that may be distinguished in state-determined systems. The state variables themselves, or the contents of integrals, are presented within rectangles, the rates of change within symbols representing values, and auxiliary variables within circles. Parameters or constants are shown as small circles on a line. The flow of material is presented by continuous arrows and the flow of information by dotted arrows. Figure 3.2 gives examples of the use of these symbols in the representation of exponential and logistic growth. In exponential growth, the amount (A) is controlled by the growth rate (GR) which is itself modified by information about the relative growth rate (RGR), set as a constant, and the present value of the amount (A). In logistic growth, A is controlled by the GR which is modified by information about the present value of A and by two constants, RGR and the carrying capacity (CC).

It is worth noting that the relational diagrams do not define the relationships explicitly. Such definition is more easily achieved by the use of mathematical equations or by computer algorithms—the unambiguous instructions which must be written to enable computers to solve the equations. However, many people find it helpful to show the main relationships symbolically as a check on the assumptions built into the mathematical expressions, and, for this reason, these diagrams are more usually drawn after the equations have been defined than before.

How does one set about the task of constructing a dynamic model of some practical problem? There is no simple answer to this question, and much depends on the primary skills of the individuals concerned with the task and, particularly, on whether the individuals who have the necessary ecological knowledge also have sufficient mathematical ability to exploit fully the capability of the mathematics. More usually, dynamic models are constructed by small teams of research scientists made up of ecologists, mathematicians and resource managers.

The most convenient start is often a relatively simple word model which can be used as a basis for the sets of system equations which will ultimately define the system. The mathematicians will attempt to formalize the relationships as quickly as possible by equations linking the state variables of the system, while the ecologists and resource managers will attempt to relate these equations to their understanding of the problem, using relational diagrams and re-interpretation of the equations as descriptions with which to compare the original word model. The whole process is an iterative one, going through several cycles of successive approximations. Sometimes, the model will draw heavily upon existing models of parts of the system, e.g. exponential or logistic growth, or will begin as a modification of an existing model. The basic mathematics acts as a convenient medium for the transfer of experience and also as a medium of communications between the mathematician and the ecologist, as was emphasized in the last chapter.

It must be stressed, however, that the set of formal equations is what is needed for the next stage of the modelling process, which is to simulate the ecological system on the computer, using the speed of the computer to allow for iterative solution of many trial values. For one thing, many of the parameters (constants) of the model will be unknown, and the computer will therefore be used to find appropriate values of these parameters by an iterative search which starts from some guessed or arbitrary values. Similarly, where functional relationships between two or more variables are assumed, both the shape and coefficients of these relationships will need to be explored in order to fit the model to observed data.

For many ecologists, the existence of special-purpose computer languages for constructing dynamic models is an added attraction. Languages such as DYNAMO, Continuous System Simulation Language (CSSL) and Continuous System Modelling Program (CSMP) have been developed to enable sophisticated use of computers by research workers without much training in advanced programming techniques, and these programming systems are intended not only to improve the communications between the research worker and the computer but also between research workers. Furthermore, in these simulation programming systems, an important feature is that all the processes and processing details may be presented in conceptual rather than computational order. The programming system itself contains a sorting routine which orders all calculations and integrations in an effective algorithm. As a result, the simulation program may be presented more clearly, and a variety of programming and conceptual errors may be detected by the system.

Nevertheless, many research workers who have both familiarity and skill in computer programming prefer to write their own simulation programs in high-level languages such as FORTRAN, BASIC or ALGOL. Of these languages, BASIC is perhaps the easiest with which to work because it is designed to be an interactive language, and this feature of the language greatly simplifies the fitting of constants and coefficients, as well as the testing of the simulation. Furthermore, the use of a general-purpose computer language necessarily avoids the limitations and rules which are inherent in the use of any special-purpose language. Compare the alternative formulations of the examples given below.

Example 3.1 – Growth of yeasts in monocultures and mixtures.

As a first example of the application of dynamic models to an ecological (more strictly, biological) problem, we will consider a relatively simple situation first described by Gause[26] who cultivated two species of yeast (*Saccharomyces cerevisiae* and *Schizosaccharomyces* 'Kephir') with known quantities of sugar. Yeasts grown under aerobic conditions with a sufficient supply of sugar, and some other substances essential to growth, consume the sugar to provide the energy for the growth of new yeast cells and for the maintenance of the yeast. The end-products (carbon dioxide and water) of the sugar broken down in the respiratory process do not pollute the environment of the yeast. If, however, the yeast grows under anaerobic conditions, one additional end-product of the respiratory process is alcohol, which may accumulate in the environment in which the yeasts are growing. This alcohol has the effect of slowing down, and ultimately stopping, the bud development of the yeasts, even when there is still enough sugar available for growth.

Gause cultivated both yeast species in monocultures, and also in mixtures, and comparison of the growth of both species in mixture with their growth in monoculture suggested that the two species had a mutual effect upon each other. His hypothesis was that the effect was due to the formation of the same waste product from both species, alcohol, and that this product affected the bud formation of both

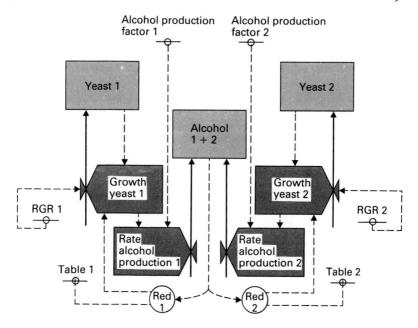

Fig. 3.3 Relational diagram for the growth and interference of two yeast species. After de Wit and Goudriaan.[17]

species. In order to test this hypothesis, we need to be able to simulate the growth of two species, independently and in mixture, under the assumption that the production of the same harmful waste product is the only cause of interaction.

A relational diagram for this example is given in Fig. 3.3. There are three state variables: i.e. the amount of the first and second yeast species and the amount of alcohol. The lines of information indicate that the growth of yeast depends on the amount of yeast, the relative growth rate of the yeast and an auxiliary variable representing a reduction factor. This reduction factor is, in turn, a function of the amount of alcohol present. The basic relations are the same for both species of yeast, although the values of parameters and the shapes of the functions may be assumed to be different. The amount of alcohol is governed by the rates of alcohol production of the two species, and these rates themselves depend on the growth rates of the species and on appropriate alcohol production factors.

De Wit and Goudriaan[17] give a CSMP program for the model, and this program is reproduced in Table 3.1. An equivalent program in

Table 3.1   CSMP program for growth and interference of two yeast species. After de Wit and Goudriaan.[17]

```
TITLE MIXED CULTURE OF YEAST
       Y1 = INTGRL (IY1, RY1)
       Y2 = INTGRL (IY2, RY2)
INCON IY1 = 0.45, IY2 = 0.45
       RY1 = RGR1 * Y1 * (1. - RED1)
       RY2 = RGR2 * Y2 * (1. - RED2)
PARAMETER RGR1 = 0.236, RGR2 = 0.049
       RED1 = AFGEN (RED1T, ALC/,MALC)
       RED2 = AFGEN (RED2T, ALC/MALC)
FUNCTION RED1T = (0., 0.), (1., 1.)
FUNCTION RED2T = (0., 0.), (1., 1.)
PARAMETER MALC = 1.5
       ALC = INTGRL (ALC, ALCP1 + ALCP2)
       ALCP1 = ALPF1 * RY1
       ALCP2 = ALPF2 * RY2
PARAMETER ALPF1 = 0.122, ALPF2 = 0.270
INCON      IALC = 0.
FINISH     ALC = LALC
           LALC = 0.99 * MALC
TIMBER     FINTIM = 150., OUTDEL 2.
PRTPLT     Y1, Y2, ALC
END
STOP
```

BASIC is given in Table 3.2, and the results derived from this program for estimated values of the parameters are presented in Table 3.3 and plotted against the experimental values derived by Gause in Fig. 3.4.

How are the estimates of the parameters derived? In this example, Gause grew the two species of yeasts in monocultures, so that it was possible, first, to obtain estimates of the rates of growth and alcohol production for each species separately, and then to compare the growth of the two species grown together with known initial amounts and at some estimated maximum alcohol concentration at which the formation of buds was totally inhibited.

Table 3.2 BASIC program for determining growth of yeasts in mixed and monocultures.

```
10 REM MIXED CULTURE OF YEASTS
20 PRINT "INITIAL AMOUNTS: Y0,Y2";
30 INPUT Y1,Y2
40 PRINT "RELATIVE GROWTH RATES: R1,R2";
50 INPUT R1,R2
60 PRINT "ALCOHOL PROD RATES: F1,F2";
70 INPUT F1,F2
80 PRINT "MAX ALCOHOL LEVEL: M";
90 INPUT M
95 PRINT
100 LET A=0
110 LET D1=0
120 LET D2=0
130 FOR I=1 TO 60
140 FOR J=1 TO 10
150 LET S1=R1*Y1*(1-D1)*0.1
160 LET S2=R2*Y2*(1-D2)*0.1
170 LET Y1=Y1+S1
180 LET Y2=Y2+S2
190 LET P1=F1*S1
200 LET P2=F2*S2
210 LET A=A+P1+P2
230 LET D1=A/M
240 LET D2=A/M
250 NEXT J
260 PRINT I,Y1,Y2,A
280 NEXT I
290 STOP
300 END
```

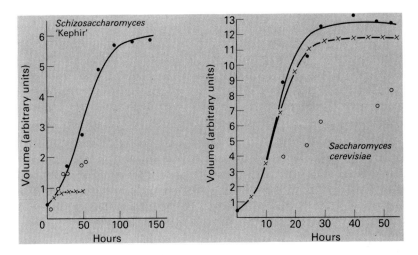

Fig. 3.4 Comparison of observed and calculated values for yeasts grown in monoculture and mixed. ○=mixed, ●=monoculture, ●——●=simulated monoculture and ×——×=simulated mixed.

Table 3.3   Observed and calculated values for growth of two species of yeasts in monocultures and mixtures.

Schizosaccharomyces 'Kephir'

| | Volume of yeast (arbitrary units) | | | |
| | Monoculture | | Mixed | |
Hours	Observed	Calculated	Observed	Calculated
0	0·45	0·45	0·45	0·45
6	—	0·60	0·291	0·59
16	1·00	0·95	0·98	0·81
24	—	1·34	1·47	0·88
29	1·70	1·64	1·46	0·89
48	2·73	3·04	1·71	0·89
53	—	3·44	1·84	0·89
72	4·87	4·72	—	—
93	5·67	5·51	—	—
117	5·80	5·86	—	—
141	5·83	5·96	—	—

Saccharomyces cerevisiae

Hours	Observed	Calculated	Observed	Calculated
0	0·45	0·45	0·45	0·45
6	0·37	1·72	0·375	1·70
16	8·87	8·18	3·99	7·56
24	10·66	11·83	4·69	10·86
29	12·50	12·46	6·15	11·47
40	13·27	12·73	—	11·75
48	12·87	12·74	7·27	11·77
53	12·70	12·74	8·30	11·77

If the relative growth rates and alcohol production factors are set at values which match the results of the two experimental mono-cultures as closely as possible, comparison of the simulated results of the mixture of species with the experimental results suggests that the actual growth of *Saccharomyces* is significantly overestimated, while that of *Schizosaccharomyces* is underestimated. We must conclude, therefore, that the two species do not interfere with each other's growth solely through the production of alcohol. It may be that either (or both) of the species produces some additional products that stimulate or inhibit the other, and we need additional biological information before the modelling of the process can proceed further.

Table 3.4 Empirical growth rates for barley and oats.

| | Relative growth rates | |
Days	Barley	Oats
0	0·4286	0·7143
7	0·1071	0·1190
14	0·0441	0·0634
21	0·0225	0·0431
28	0·0064	0·0242
35	−0·0036	0·0511
42	−0·0065	0·0491

Example 3.2 – Barley and oats growing in competition.

For the second example of dynamic models, we will use the Lotka-Volterra equations:

$$\frac{dx}{dt} = x(a - bx - cy)$$

$$\frac{dy}{dt} = y(e - fx - gy)$$

where a, b, c, e, f and g are positive constants to represent the competition between plant species in a mixed stand. In contrast to the first example, the number of plants of each species in a mixed stand of an agricultural crop is determined at the time of planting. In our model, the competition between each crop species will be expressed in terms of 'relative space', a dimensionless-variable which characterizes the effect of crowding on available root and foliage space, nutrients, sunlight, and associated factors. The actual production of dry matter can then be obtained from the product of 'relative space' and the maximum possible yield for dense monocultures. All three quantities are therefore functions of time.

In our example, we will simulate the growth and competition of barley and oats planted as a mixed stand. The differential equations, in which r_b represents relative space for barley and r_o represents relative space for oats, are as follows:

$$\frac{dr_b}{dt} = G_b(r_b - r_b^2 - r_b r_o)$$

$$\frac{dr_o}{dt} = G_o(r_o - r_o^2 - r_b r_o)$$

Table 3.5   CSMP program for barley and oats growing in competition.

```
TITLE BARLEY AND OATS -- COMPETING FOR THE SAME SPACE
INITIAL
       RSO1 = 3.0 * DO
       RSB1 = 9.0 * DB
DYNAMIC
*      EQUATIONS FOR OATS
       RSO = INTGRL (RSO1, RATEO)
       RATEO = RSO * (1. - RSO - RSB) * AFGEN (RGRO, TIME)
*      EQUATIONS FOR BARLEY
       RSB = INTGRL (RSB1, RATEB)
       RATEB = RSB * (1. - RSO - RSB) * AFGEN (RGRB, TIME)
FUNCTION RGRB = (0.0, 0.4286), ....
               (7.0, 0.1071), ....
               ..................
               ..................
               (42.0, -0.0065),...
FUNCTION RGRO = (0.0, 0.7143), ....
               (7.0, 0.1190), ....
               ..................
               ..................
               (42.0, 0.0491), ...
METHOD MILNE
PRINT RBS, RSO, DELT
TIMBER    FINTIM = 42.0, PRDEL = 7.0
*      DENSITY OF SOWING ON ROW/CM
PARAMETER  DB = 0.02, DO = 0.02
END
STOP
```

In order to simulate the competition with time, we will employ a device which is frequently used in such models, i.e. to regard the coefficients G_b and G_o as empirical functions of time given in Table 3.4. Both functions represent relative plant growth rates in the absence of competition which have been determined from experimental plantings with very low seed densities (Baeumer and de Wit[3]).

Table 3.6 BASIC program for barley and oats growing in competition.

```
10 REM COMPETITION BETWEEN BARLEY AND OATS
20 PRINT "SOWING DENSITIES IN ROW/CM";
30 INPUT D1,D2
35 PRINT\PRINT\PRINT
40 LET S1=3.0*D1
50 LET S2=9.0*D2
60 DIM P(7),R1(7),R2(7)
70 FOR I=1 TO 7
80 READ P(I),R1(I),R2(I)
90 NEXT I
100 LET T=0
110 FOR I=1 TO 42
120 FOR J=1 TO 100
130 FOR K=2 TO 7
140 IF I>P(K) THEN 180
150 LET G1=R1(K-1)+(T-P(K-1))*(R1(K)-R1(K-1))/7
160 LET G2=R2(K-1)+(T-P(K-1))*(R2(K)-R2(K-1))/7
170 GO TO 190
180 NEXT K
190 LET S3=S1*(1-S1-S2)*G1*0.01
200 LET S4=S2*(1-S2-S1)*G2*0.01
205 LET S1=S1+S3
207 LET S2=S2+S4
210 LET T=T+0.01
220 NEXT J
230 PRINT I,G1,G2,S1,S2
240 NEXT I
250 STOP
251 DATA 0,0.7143,0.4286
252 DATA 7,0.1190,0.1071
253 DATA 14,0.0634,0.0441
254 DATA 21,0.0431,0.0225
255 DATA 28,0.0242,0.0064
256 DATA 35,0.0511,-0.0036
257 DATA 42,0.0491,-0.0065
300 END
```

Empirical functions like these relative growth rates are easily modelled by means of the special-purpose modelling languages— indeed the ability to incorporate such functions is one of the main advantages of these languages. Brennan et al.[7] give a CSMP program for this model, and part of this program is presented in Table 3.5. In this program, the functions are derived from linear interpolation of the tabulated points, but CSMP also allows for quadratic interpolation between the data points if this level of accuracy is justified.

An equivalent BASIC program is given in Table 3.6, with the empirical functions also approximated by linear interpolation. The output from this program is listed in Table 3.7 and plotted in Fig. 3.5. Because the barley grows rapidly at first and then levels off, and the reverse is true for oats, when the two species are grown together, a disproportionate share of the available space is occupied by the barley

Table 3.7   Results from BASIC program of Table 3.6.

```
SOWING DENSITIES IN ROW/CM? 0.02,0.02
```

1	.6301076	.3831307	.09683289	.2403713
2	.5450647	.3372021	.1387513	.299701
3	.4606219	.2912736	.179684	.3522976
4	.3749791	.2453451	.214875	.3952106
5	.2899363	.1994165	.2421353	.4280727
6	.2048936	.153488	.2612913	.4519494
7	.1198508	.1075595	.2730962	.468282
8	.1111366	.09819005	.28102	.4803851
9	.1031938	.08919006	.2880248	.4908378
10	.09525091	.08019006	.2941902	.4997922
11	.08730806	.07119007	.2995939	.5073929
12	.07936521	.06219008	.3043093	.5137725
13	.07142236	.05319009	.3083993	.5190492
14	.06347951	.0441901	.3119251	.5233257
15	.06052903	.04104518	.3150599	.5269338
16	.05762904	.03795947	.3179564	.5301679
17	.05472904	.0348737 6	.3206301	.5330534
18	.05182904	.03178805	.3230954	.5356136
19	.04892904	.02870233	.3253657	.5378698
20	.04602905	.02561662	.3274531	.5398408
21	.04312905	.02253091	.3293688	.5415438
22	.04042705	.02022304	.331127	.5438206
23	.03772705	.01792304	.33274	.5443099
24	.03502706	.01562305	.3342153	.5454211
25	.03232706	.01332305	.3355593	.5463624
26	.02962706	.01102305	.336778	.547141
27	.02692706	8.723054E-3	.3378765	.5477632
28	.02422707	6.423056E-3	.3388594	.5482341
29	.02800433	4.985750E-3	.3398535	.5485846
30	.03184718	3.557180E-3	.3409836	.5488441
31	.03569004	2.128610E-3	.3422469	.5490146
32	.03953289	7.060401E-4	.3436406	.5490982
33	.04337584	-7.285637E-4	.3451618	.5490972
34	.04721878	-2.157168E-3	.3468074	.5498141
35	.05106173	-3.585772E-3	.3485743	.5483512
36	.05081713	-4.010169E-3	.3503368	.548639 1
37	.0505314	-4.424464E-3	.3521704	.5484076
38	.05024568	-4.838759E-3	.3539257	.5481573
39	.04995996	-5.253055E-3	.3556533	.5478888
40	.04967424	-5.667350E-3	.3573538	.5476027
41	.04938852	-6.081645E-3	.3590278	.5472996
42	.0491028	-6.495940E-3	.360676	.5469798

at an early stage, and, by the time the oats begin to grow, there is not
enough space to accommodate them.

Example 3.3 – Upstream migration of salmon.

The third example of dynamic models which we will consider is the
simulation of upstream migration of salmon discussed by Radford.[65]

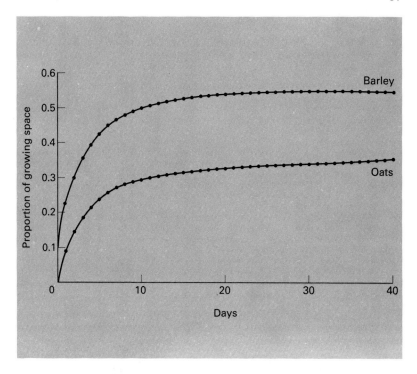

Fig. 3.5 Growth of oats and barley from initial spacing of 0·02 cm.

The fundamental components of the system are summarized in the relational diagram of Fig. 3.6. The salmon migration is postulated to depend on the quantity of fresh water in the estuary rather than on the water level at the weir at which the rate of salmon migration and the water level are measured. The upper continuous line of the diagram indicates the passage of salmon from the sea to the head of the river, via the estuary and the weir. The lower continuous line indicates the passage of water from the head of the river to the sea. The three state variables of interest are the number of salmon in the estuary, the number of salmon at the head of the river, and the number of salmon waiting for lower flows. The fourth state variable is the amount of fresh water in the estuary. The model depends upon several constants, notably the height of water at the weir, the threshold water level at high flows, the total number of salmon and a delay factor for the rate of mixing of fresh water with sea water.

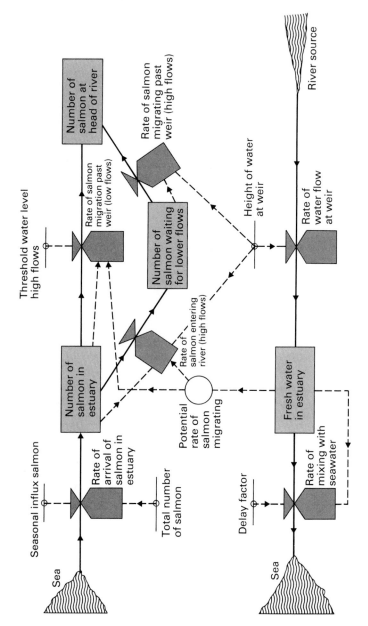

Fig. 3.6   Relational diagram for upstream salmon migration. After Radford.[65]

Table 3.8   Part of DYNAMO program simulating upstream migration of salmon.

NOTE	UPSTREAM MIGRATION OF SALMON	
NOTE	PROGRAMMED FOR AN IBM S/360 COMPUTER	
L	NSE.K = NSE.J + DT * (RASE.JK-RSMW.JK-RSER.JK)	NO OF SALMON IN ESTUARY
R	RMS.KL = FWE.K/DELAY	RATE OF MIXING OF SEAWATER
R	RASE.KL = TNSTY * SIS.K/100	RATE OF ARRIVAL OF SALMON
A	SIS.K = TABHL (TDRSI, TIME, 15, 201, 31)	SEASONAL INFLUX SALMON
T	TDRSI = .083/.131/.129/.133/.323/1.370/1.108	TABLE SIS
A	PRSM.K = TABHL (TPRSM, FWE, K, .6, 1.6, 1.)	POTENTIAL RATE MIGRATION
T	TPRSM = 0.0/0.95	TABLE PRSM
L	FWE.K = FWE.J + DT * (RWFW.JK - RMS.JK)	FRESHWATER IN ESTUARY
.	...	...
.	...	...

The model also requires information on the seasonal influx of salmon and the potential rate of salmon migrating. In the absence of experimental evidence for a particular weir, a useful understanding of the operation of the system can be gained by reasonable estimates of these relationships. Table 3.8 shows part of the DYNAMO program suggested by Radford[65] in which such estimates are incorporated as simple tables.

The number of salmon in the estuary at a given time (NSE.K) is defined in terms of the numbers that were present (NSE.J) one small interval of time (DT) before, and the rates of arrival (RASE.JK) and departure (RSMW.JK, RSER.JK) that operated over that interval. It is assumed that the value of DT chosen will make it reasonable to assume that these rates are constant over that short period of time. A similar form of equation is written for each of the state variables, and, in each case, the program indicates the rates which must be included. The rate equations can also be written down as functions of the state variables and the arbitrary functions introduced as assumptions. For example, the rate of mixing of the estuary water with the sea over a small interval of time (RMS.KL) depends upon the quantity of fresh water in the estuary at time K (FWE.K) divided by the delay constant. The rate of arrival of salmon in the estuary depends upon the seasonal influx of salmon into the estuary from a function (TABHL TDRSI, TIME, 15, 201, 31). The DYNAMO notation indicates that the data depend upon TIME according to some arbitrary function tabulated at points 31 days apart, starting at day 15 and finishing at day 201. Linear interpolation is assumed between the points of the table, and the first value is taken to apply if TIME is less than 15 days, and the last value if TIME is greater than 201 days.

As with CSMP, there will, in general, be one line of coding for each symbol in the flow chart and the order in which these lines are written is irrelevant as the DYNAMO system is able to sort the equations into the proper computational sequence. There is, however, no difficulty in translating the relationships into a general-purpose language like BASIC, with the tabulated values translated into curves fitted by orthogonal polynomials. The example illustrates the use of a dynamic model of a relatively complex ecological problem, for which many of the necessary data may not be readily available. The advantage of using such a model is that the sensitivity of the system to changes in the assumed parameters and functions can be explored in some detail. The emphasis of future research can then be placed on the direct measurement of those parameters which are shown to be important in determining the way in which the system operates.

Example 3.4 – Yield table of Sitka spruce.

As a fourth example, we will look at an empirical model for constructing forest yield tables derived by Christie,[11] which illustrates an early application of dynamic models—indeed, an application of the technique before it became widely known as dynamic modelling. The example is particularly useful in illustrating that the basic ideas are not new, and also shows a somewhat extreme degree of empiricism.

A yield table is the forester's traditional way of showing the progressive development of a stand of trees at periodic intervals covering the greater part of its useful life on a given site. Such a table usually includes the average diameter and height of the trees, their basal area (i.e. the sum of the cross-sectional areas of the trees at breast height) and their volume.

Yield tables have been produced from the early days of scientific forestry, using largely graphical methods of fitting curves to data derived from measured sample plots. As new statistical techniques were developed, attempts were made to improve the characterization of forest stands by means of curve fitting routines. However, as the various parts of the yield table are mutually dependent, it is important to ensure that the relationships of the fitted curves reflect the interdependence of the variables. In Britain, a principal feature of this

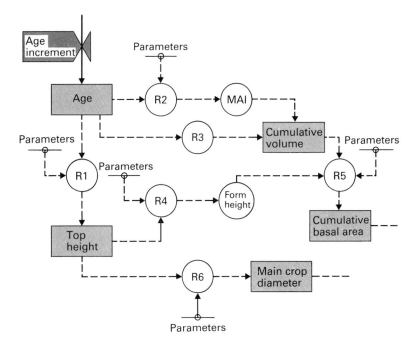

Fig. 3.7 Relational diagram for simple yield table.

process was the use of a 'master table' in which relationships for all the important crop characteristics are constructed with respect to 'top height' (i.e. the average height of the 100 trees of largest diameter per hectare). The yield tables corresponding to quality or site classes are then derived from the master table relationship by substituting age for height from an ancillary top height/age relationship.

Figure 3.7 gives a relational diagram for a simple yield table, and Table 3.9 gives a BASIC program for the calculations of the yield table for Sitka spruce, *Picea sitchensis* (Bong.) Carr. For each age derived from a series of increments, the mean annual increment and the top height are calculated from complex relationships depending upon given parameters. The state variable of cumulative volume is derived from age and mean annual increment by a further relationship, and both form height and main crop diameter are similarly derived from the top height. Finally, the cumulative basal area is derived from yet another set of parameters relating cumulative volume and form height.

It should be emphasized that Table 3.9 forms only part of the full model that would be required for a yield table. For most purposes, the

Table 3.9   BASIC program for yield of Sitka spruce.

```
LIST
   10 REM SHORT YIELD TABLE FOR SITKA SPRUCE
   20 FOR A=20 TO 80 STEP 5
   30 LET H=((5.6580760E-14*A-2.3814098E-11)*A+3.7952208E-9)*A
   31 LET H=(((H-2.9563399E-7)*A+1.3048857E-5)*A-4.7671981E-4)*A
   32 LET H=((H+1.5601327E-2)*A+1.9162451E-1)*A+9.8037440E-2
   40 LET M=((-4.9693779E-13*A+2.0023260E-10)*A-3.4496064E-8)*A
   41 LET M=(((M+3.2853312E-6)*A-1.8590905E-4)*A+6.2650400E-3)*A
   42 LET M=((M-1.2572190E-1)*A+1.8334816)*A-1.1562462E+1
   50 LET V=M*A
   60 LET F=-0.600441+0.468491*H-0.003619*H*H
   70 LET G=V/F
   80 LET D=1.673873+0.087942*H-0.000719*H*H
   81 LET D=EXP(D)
   90 PRINT A,INT(H*10+0.5)/10,INT(G*10+0.5)/10,INT(V+0.5),INT(D*10+0.5)@
  100 NEXT A
  110 STOP
  120 END

READY
```

```
RUN

   20        7.7          26.8        75       10.1
   25       10.2          38         144       12.1
   30       12.7          47.9       228       14.5
   35       15.1          56.6       320       17.1
   40       17.4          64.1       413       19.8
   45       19.4          70.5       503       22.4
   50       21.2          76         585       24.9
   55       22.7          80.5       657       27.1
   60       23.9          84.3       720       29
   65       25            87.5       774       30.6
   70       25.9          90.3       821       32.1
   75       26.7          92.7       864       33.4
   80       27.3          94.8       901       34.5

READY
```

forester would need to include relationships involving the numbers of trees left after successive thinnings and to distinguish between the standing crop of trees and those removed at various stages in the life of the stand as thinnings. Note, however, that, even in this very simple yield table, there are six functional relationships and five sets of parameters. Furthermore, although the model is non-linear, it does not have any feed-back loops, at least in this simple form.

How are the relationships determined for a model of this type? The traditional method of derivation is to fit curves of various kinds through observed data from permanent sample plots which have been measured at regular intervals, using the statistical technique of orthogonal polynomials constrained to pass through a small number of theoretically 'fixed' points. In some cases, no single curve could be fitted, even with the high order polynomials necessary to avoid the sinuosity inherent in this kind of function, and pseudo-spline curves were used. The point to be stressed, however, is that this process of fitting many separate relationships independently does not necessarily lead to consistent and compatible relationships, and the purpose of the model is to test this internal consistency. There are also some important problems about the sensitivity of the relationships to small changes in the shapes of the functions and the estimates of the parameters, and these problems will be discussed in more detail when we consider the relative advantages and disadvantages of dynamic models as a family of models.

Stand yield models of this kind are widely used as practical tools of forest research and management. The idea of representing the growth of a population of trees by a brief tabular statement is something which foresters have taken for granted for a very long time, although, in related disciplines, similar approaches to describing the development of biological populations are regarded as innovations. Hamilton and Christie[31] describe the use of yield models similar to (but more complex than) that of Fig. 3.7 and Table 3.9 for production forecasting, evaluating alternative treatments, valuation of stands, and yield control. Such models are generally easier to construct and to use than models based on the growth of individual trees which, at present, are relatively unreliable in predicting yield in terms of unit area.

Having looked at some simple examples of dynamic models, we can now discuss the relative advantages and disadvantages of this family of mathematical models. Clearly, such models have an intuitive appeal to many ecologists, especially if they also have some reasonable mathematical background. The formulation of the models allows for considerable freedom from constraints and assumptions, and allows for the introduction of the non-linearity and feed-back which are apparently characteristic of ecological systems. The ecologist is able to mirror or mimic the behaviour of the system as he understands it, and gain some useful insight into the behaviour of the system as a result of changes in the parameters and driving variables. The power of computers to make large numbers of exact but small computations also enables the ecologist to replace the analytical technique of integration by the less accurate, but easier, methods of difference equations. Furthermore, even where the values of parameters are unknown, relatively simple techniques exist to provide approximations for these parameters by sequential estimates, or even to use interpolations from tabulated functions. In particularly favourable cases, it may even be possible to test various hypotheses about parameters or functions.

The lack of a formal structure for the models, and the freedom from constraints, can also be disadvantageous. For one thing, the behaviour of even quite simple dynamic models may be very difficult to predict. It requires only one non-linearity and two feed-back loops to create a model system whose behaviour will almost certainly be counter-intuitive. As a result, it is desperately easy to construct models whose behaviour, even within the practical limits of the input variables, is unstable or inconsistent with reality. Even more difficult, determination of the way in which such systems behave will frequently require extensive and sophisticated experimentation. For example, it is nearly always necessary to test the behaviour of the model in relation to the interactions of changes of two or more input variables, and seldom, if ever, sufficient to test the response to changes in one variable at a time.

The inability to predict the behaviour of dynamic models severely limits their value in the development of further theory. As we will see, some of the other families of models behave in more predictable ways so that, in exchange for more clearly established assumptions, the responses of the model to changes can be more readily deduced. Admittedly, much of the difficulty in constructing mathematical models is then focussed on the testing of the basic assumptions necessary for the use of the model, but this testing will usually be easier, and mathematically more rigorous, than the search for the complex modes of behaviour and discontinuities of dynamic models.

Perhaps the most important disadvantage of dynamic models, however, is the uncertainty of being able to estimate the values of the basic parameters, especially where there are many such parameters to be estimated. Although, as we have indicated, methods are available for deriving estimates of parameters by successive approximation, such methods are usually time-consuming and may be tedious, even on a computer. Furthermore, it is not always possible to arrive at convergent estimates for even relatively simple models, and many of the other families of mathematical models have been specially designed to simplify the estimation of the basic parameters, even at the expense of apparent 'reality'. The disadvantage is frequently compounded by the lack of parsimony in the number of relationships, variables, and parameters that is often a feature of dynamic models because of the otherwise praiseworthy desire to mimic reality as closely as possible. For the scientist, it is always necessary to work towards the simplest possible model, so that the entities and relationships are not multiplied needlessly, but this simplicity is not easy to balance against the flexibility of the dynamic models.

Because of their nature and the underlying mathematics of their structure, dynamic models are heavily oriented towards deterministic solutions. Admittedly, stochastic relationships can often be incorporated in such models, but sometimes only with difficulty. As a result, the models of this family do not usually reflect the inherent variability of ecological and biological systems. As will be argued later, it is particularly important to be able to model the variability of ecological systems as well as their average tendencies—indeed the stability of such systems may depend upon this variability.

Finally, although dynamic models are usually highly dependent upon the use of computers, the construction of the models and their exploration may require very large amounts of computing time, and, especially if the special-purpose simulation languages are employed, large computer facilities and configurations. Such models are, therefore, frequently expensive and very demanding of resources that may be difficult to acquire, particularly in developing countries or in small research organizations.

To summarize, therefore, dynamic models may well be helpful in the early stages of the systems analysis of a complex ecological problem by concentrating attention on the basic relationships underlying the system and by defining the variables and sub-systems that the investigator believes to be critical. In the later stages of the analysis, however, it will usually be preferable to switch the main effort to one of the other families of models. It is, however, precisely for this reason that systems analysis explicitly defines a phase of developing alternative solutions to the problem.

The models of this chapter reflect the difficulties of parameter estimation to varying degrees. Example 3.1 is a classical example for which the difficulty is minimized. The model is based on simple and well-defined relationships and the parameters are estimated from the experimental results on the monocultures. The results for the mixed culture follow logically from these estimates. Example 3.2 introduces a further complication, in that, although the basic relationships are still clearly defined, the estimation of the parameters is now more complex. The estimation of the relative growth rates is avoided by linear interpolation between fixed points of empirical functions of the growth rates of the two species with time. How sensitive is the model to changes in the shapes of these functions? This sensitivity could clearly be tested, but there is an infinity of possible combinations of shapes for the two species, and the test is not entirely unambiguous.

Example 3.3 stretches the complications even further, with more parameters to be estimated, less simple relationships and increased dependence on arbitrary functions. The testing of the sensitivity of this model to changes in the assumptions has now become a major pre-occupation, occupying more time than any other part of the modelling process. The level of empiricism is complete in Example 3.4. This model is wholly dependent upon a series of arbitrary functions, admittedly derived from curve-fitting procedures of varying kinds based on actual data, but the model is now wholly reduced to a way of testing the internal consistency of the relationships incorporated.

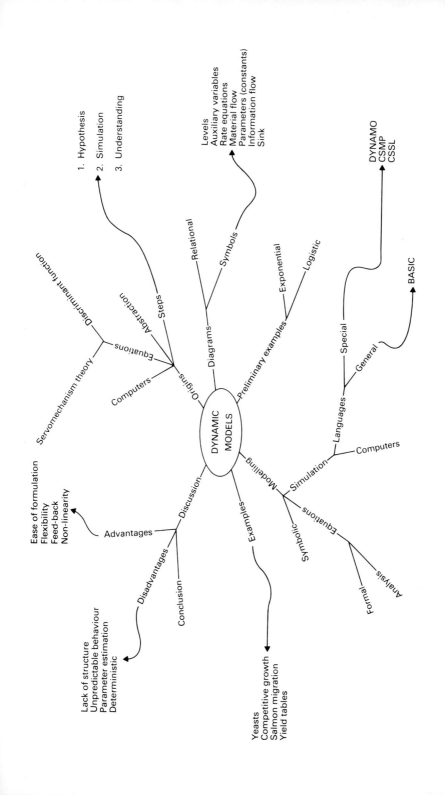

4

Matrix Models

In the family of dynamic models described in the previous chapter, we have examined one possible approach to the modelling of ecological systems. Such models offer almost complete freedom to the investigator for the expression of those elements which he considers to be essential to the understanding of the underlying relationships between those variables and entities which he has identified in the description of the system. The models usually strive for 'reality'—a recognizable analogy between the mathematics and the physical, chemical, or biological processes—sometimes at the expense of mathematical elegance or convenience. The price paid for the 'reality' is frequently a necessity to multiply entities to account for relatively small variations in the behaviour of the system, or difficulty in deriving unbiased or valid estimates of the model parameters. Matrix models represent one family of models in which 'reality' is sacrificed to some extent in order to gain the advantages of the particular mathematical properties of the formulation. The deductive logic of pure mathematics then enables the modeller to examine the consequences of his assumptions without the need for time-consuming 'experimentation' on the model, although computers may still be required for some of the computations.

The term 'matrix' is used by mathematicians to describe a rectangular array of numbers, and the matrix

$$\mathbf{A} = \begin{bmatrix} 1 & 0 & 3 & -6 \\ 2 & 4 & 8 & 16 \\ 1 & 3 & 2 & -5 \end{bmatrix}$$

is a matrix of three rows and four columns, or a 3×4 matrix. Each of the twelve numbers in the matrix is termed an *element* of the matrix, and if the matrix as a whole is **A**, then the element a_{ij} is the element in the ith row and jth column of **A**. Thus, in symbols, the above matrix is:

$$\mathbf{A} = \begin{bmatrix} a_{11} & a_{12} & a_{13} & a_{14} \\ a_{21} & a_{22} & a_{23} & a_{24} \\ a_{31} & a_{32} & a_{33} & a_{34} \end{bmatrix}$$

Note the convention that a capital, bold-faced letter is always used to represent a matrix and that the corresponding lower-case letter, with appropriate subscripts, is used to represent any element of that matrix. Some special kinds of matrices occur frequently in the mathematical manipulations of these arrays of numbers, known as 'matrix algebra' or 'linear algebra'. For example, a matrix with the same number of rows and columns is a *square matrix*, and three forms of square matrices are given special names.

$$\mathbf{I} = \begin{bmatrix} 1 & 0 \\ 0 & 1 \end{bmatrix} \qquad \mathbf{O} = \begin{bmatrix} 0 & 0 \\ 0 & 0 \end{bmatrix} \qquad \mathbf{A} = \begin{bmatrix} 3 & -2 \\ -2 & 5 \end{bmatrix}$$

I being a *unit matrix*, **O** the *null matrix*, and **A** a *symmetric matrix*. The elements of the first row and first column, second row and second column, etc., dominate the matrix and are known as the *principal diagonal* of the matrix. In the *unit*, or *identity*, *matrix*, the principal diagonal consists of 1's and all the other elements are zeros. In a *symmetric matrix*, the elements of the principal diagonal can take any values, but the off-diagonal elements are related such that $a_{ij} = a_{ji}$.

Finally, matrices with only one row or one column:

$$\mathbf{a} = \begin{bmatrix} 1 & 3 & 2 \end{bmatrix} \quad \text{or} \quad \mathbf{b} = \begin{bmatrix} 2 \\ 4 \\ 6 \end{bmatrix}$$

are known as *row* and *column vectors* respectively. Conventionally, they are represented by bold-faced lower-case letters. A matrix with only one element

$$\mathbf{D} = \begin{bmatrix} 3 \end{bmatrix}$$

is known as a *scalar*.

The advantage of expressing arrays of numbers as matrices is that these arrays can be manipulated in ways which are analogous to the manipulation of ordinary numbers, or scalars. For example, the addition or subtraction of two matrices corresponds to the addition or subtraction of each corresponding element of the matrix. Multiplication and division of matrices are more complex, but remain unambiguous mathematical operations. Matrix algebra is one of the most important developments of modern mathematics, and the ecologist who intends to use systems analysis is recommended to study one or more of the available introductory text-books; for example, see Searle,[71] Anton[2] and Rorres and Anton,[66] to gain familiarity with the notation and operations of this branch of mathematics. Square matrices have an important property in that, for any square matrix, there are eigenvalues and eigenvectors which satisfy the equation:

$$\mathbf{Av} = \lambda \mathbf{v}$$

where $\mathbf{A}$ is a square matrix, $\mathbf{v}$ is a column vector and λ is a scalar. In general, if $\mathbf{A}$ is a $n \times n$ matrix, then n values of λ can be calculated, although some of these eigenvalues may be repeated, negative or imaginary. For every eigenvalue λ, there is an associated eigenvector $\mathbf{v}$, and, as we shall see later in this chapter, these characteristics are valuable in summarizing particular features of the original matrix. Various methods exist for the calculation of eigenvalues and eigenvectors, and we will encounter some of these methods when we come to consider examples of matrix models.

One of the earliest forms of matrix model was developed by Lewis[50] and Leslie[49] as a deterministic model predicting the future age structure of a population of female animals from the present known age structure and assumed rates of survival and fecundity. The population is first divided into $n + 1$ (i.e. 0, 1, 2, 3, ..., n) equal age groups so that the oldest group possible, or the age group in which all the animals surviving die, is n. The model is then represented by the matrix equation:

$$
\begin{bmatrix}
f_0 & f_1 & f_2 & \cdots & f_{n-1} & f_n \\
p_0 & 0 & 0 & \cdots & 0 & 0 \\
0 & p_1 & 0 & \cdots & 0 & 0 \\
0 & 0 & p_2 & \cdots & 0 & 0 \\
\cdot & \cdot & \cdot & & \cdot & \cdot \\
\cdot & \cdot & \cdot & & \cdot & \cdot \\
\cdot & \cdot & \cdot & & \cdot & \cdot \\
0 & 0 & 0 & \cdots & p_{n-1} & 0
\end{bmatrix}
\begin{bmatrix}
n_{t,0} \\
n_{t,1} \\
n_{t,2} \\
n_{t,3} \\
\cdot \\
\cdot \\
\cdot \\
n_{t,n}
\end{bmatrix}
=
\begin{bmatrix}
n_{t+1,0} \\
n_{t+1,1} \\
n_{t+1,2} \\
n_{t+1,3} \\
\cdot \\
\cdot \\
\cdot \\
n_{t+1,n}
\end{bmatrix}
$$

In this equation, the numbers of animals in the various age classes at time $t + 1$ are obtained by multiplying the numbers of animals in these age classes at time t by a matrix expressing the appropriate fecundity and survival rates for each age class. The f_i ($i=0$, 1, 2, ..., n) represent the reproduction by the female in the ith age group, and the p_i ($i=0$, 1, 2, ..., $n-1$) represent the probabilities that a female in the ith age group will be alive in the $(i+1)$ age group.

What is perhaps less obvious is that the behaviour of this model can be predicted by formal analysis of the matrix $\mathbf{A}$ in the matrix equation:

$$\mathbf{A}\mathbf{a}_t = \mathbf{a}_{t+1}$$

where $\mathbf{a}_t$ is the column vector representing the population age structure at time t, and $\mathbf{a}_{t+1}$ is a column vector representing the age structure at time $t+1$. First, by a simple extension of the equation, we can predict the numbers of animals in the age classes after k periods of time by repeating the multiplication, so that

$$\mathbf{a}_{t+k} = \mathbf{A}^k \mathbf{a}_t$$

Second, as the matrix $\mathbf{A}$ is square with $n+1$ rows and columns, there are $n+1$ possible eigenvalues and eigenvectors. The elements of $\mathbf{A}$ are either positive or zero, because neither the f_i nor the p_i can take negative values, and we can show that, in this case, the largest eigenvalue and the corresponding eigenvector is ecologically meaningful.

A simple example will help to show the significance of the eigenvalue and eigenvector, and the example given by Williamson[97] is one of the simplest models with which to start. He gives the model:

$$\begin{bmatrix} 0 & 9 & 12 \\ \frac{1}{3} & 0 & 0 \\ 0 & \frac{1}{2} & 0 \end{bmatrix} \begin{bmatrix} 0 \\ 0 \\ 1 \end{bmatrix} = \begin{bmatrix} n_{t+1,0} \\ n_{t+1,1} \\ n_{t+1,2} \end{bmatrix}$$

This model starts with an initial population of one old female, represented by the column vector. After one period of time, there will be 12 young females, i.e.

$$\begin{bmatrix} 0 & 9 & 12 \\ \frac{1}{3} & 0 & 0 \\ 0 & \frac{1}{2} & 0 \end{bmatrix} \begin{bmatrix} 0 \\ 0 \\ 1 \end{bmatrix} = \begin{bmatrix} 12 \\ 0 \\ 0 \end{bmatrix}$$

Repeated application of the model gives the following predictions, in which the latest population is pre-multiplied by the fecundity and survival rates.

$$\begin{bmatrix} 0 & 9 & 12 \\ \frac{1}{3} & 0 & 0 \\ 0 & \frac{1}{2} & 0 \end{bmatrix} \begin{bmatrix} 12 \\ 0 \\ 0 \end{bmatrix} = \begin{bmatrix} 0 \\ 4 \\ 0 \end{bmatrix}$$

$$\begin{bmatrix} 0 & 9 & 12 \\ \frac{1}{3} & 0 & 0 \\ 0 & \frac{1}{2} & 0 \end{bmatrix} \begin{bmatrix} 0 \\ 4 \\ 0 \end{bmatrix} = \begin{bmatrix} 36 \\ 0 \\ 2 \end{bmatrix}$$

$$\begin{bmatrix} 0 & 9 & 12 \\ \frac{1}{3} & 0 & 0 \\ 0 & \frac{1}{2} & 0 \end{bmatrix} \begin{bmatrix} 36 \\ 0 \\ 2 \end{bmatrix} = \begin{bmatrix} 24 \\ 12 \\ 0 \end{bmatrix} \quad \text{etc.}$$

Each old animal produces an average of 12 young before dying; each middle-aged animal produces an average of 9 young before either dying or becoming one age-period older, with equal probability. Young animals produce no young and have a probability of $\frac{1}{3}$ of reaching the middle age group. The numbers of animals predicted for each age group are plotted on a logarithmic scale for the first 20 time periods in Fig. 4.1. After some initial instability, the predicted numbers increase exponentially and the numbers of young, middle-aged and old animals maintain a constant ratio to each other.

The dominant eigenvalue and eigenvector of a matrix can be calculated by a relatively simple method of successive approximations. The dominant eigenvalue gives the rate at which the population size is increased, and in our case this eigenvalue is 2, indicating that, during each period of time, the population size is doubled. More generally, if the dominant eigenvalue is λ, then

$$\mathbf{Av} = \lambda \mathbf{v}$$

where $\mathbf{v}$ is the stable population structure, measured by proportions rather than by actual numbers. If the logarithm of population size is plotted against time, the slope of the line after the stable population structure has been reached is equivalent to $\ln \lambda$, the intrinsic rate of natural increase. The dominant eigenvalue can also be used to estimate the number of individuals that can be removed from the population, to bring it back to its initial population size, by the equation:

$$H = 100 \left(\frac{\lambda - 1}{\lambda} \right)$$

where H is expressed as a percentage of the total population.

Similarly, the dominant eigenvector indicates the stable structure of the population. In our example, this vector is:

$$\begin{bmatrix} 24 \\ 4 \\ 1 \end{bmatrix}$$

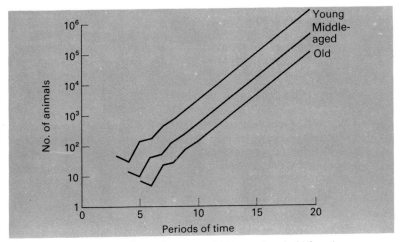

Fig. 4.1   Predicted numbers of young, middle-aged and old females.

and shows the proportions of young, middle-aged and old animals in the stable population.

This example illustrates the basic reason for using a more restricted formulation of the mathematical model, in that relatively simple calculation indicates some of the principal properties of the model. Our example suffers from the same disadvantages as the deterministic exponential model of population growth in that it assumes that the population size will continue to increase. A more realistic model can, however, be readily formulated by making all the elements of the matrix functions of some property of population size. We will see an example of this modification later in this chapter.

Predator-prey systems, which sometimes show marked oscillations, can also be encompassed by matrix models, by a relatively simple exploitation of the techniques for relating population size and survival. Seasonal and random environmental changes and the effects of time-lags may similarly be incorporated, although the models necessarily become increasingly complex in formulation.

There are many developments of the basic matrix model which we have outlined above. All of these developments represent modifications or additions to the elements of the matrix, and one of the simplest of these is investigation of the effects of harvesting different parts of the population. Lefkovitch,[48] for example, gives a useful introduction to the mathematics of harvesting models which he derived from his original models for immigration. Similarly, the model can be readily extended to include both sexes and Williamson[96] gives the simple case of a population divided into only three age classes.

$$
\begin{bmatrix}
0 & f_{m0} & 0 & f_{m1} & 0 & f_{m2} \\
0 & f_{f0} & 0 & f_{f1} & 0 & f_{f2} \\
p_{m0} & 0 & 0 & 0 & 0 & 0 \\
0 & p_{f0} & 0 & 0 & 0 & 0 \\
0 & 0 & p_{m1} & 0 & 0 & 0 \\
0 & 0 & 0 & p_{f1} & 0 & 0
\end{bmatrix}
\begin{bmatrix}
n_{t,m0} \\
n_{t,f0} \\
n_{t,m1} \\
n_{t,f1} \\
n_{t,m2} \\
n_{t,f2}
\end{bmatrix}
=
\begin{bmatrix}
n_{t+1,m0} \\
n_{t+1,f0} \\
n_{t+1,m1} \\
n_{t+1,f1} \\
n_{t+1,m2} \\
n_{t+1,f2}
\end{bmatrix}
$$

where f_{mi} and f_{fi} are the numbers of males and females respectively produced by a female in the ith age class during one period of time, p_{mi} and p_{fi} are the probabilities that a male and female respectively will survive from one period to the next, and $n_{t,mi}$ and $n_{t,fi}$ are the numbers of males and females respectively in the ith age group at time t.

Perhaps the best-known modifications of matrix models, however, have been concerned with consideration of size structures or discrete stages within the population. For example, Usher[78,79,80,81] has used such models to investigate the management and harvesting of forests, where trees are classified by size as well as age. In contrast, Lefkovitch[46,47,48] has applied matrix models to insect pests of stored products, where the structure of the insect populations is defined by the development stages of the insect life cycle.

Dynamic processes such as the cycling of nutrients and the flow of energy in ecosystems can also be modelled by the use of matrices, exploiting the natural compartmentation of the ecosystem into its species composition or into its trophic levels. Losses from the ecosystem are assumed to be the difference between input and the sum of the output from, and storage in, any one compartment. Finally, we will explore an extension of the concept of matrix models when we come to consider Markov models in the chapter on stochastic models. Markov models are distinguished by the fact that the sum of the elements in each column is unity.

Example 4.1 – Survival and fecundity of the blue whale.

Our first example of the practical application of matrix models is similar to Leslie's basic model, and was described by Usher[82] from data on the blue whale (*Balaenoptera musculus*) provided by Laws[45] and Ehrenfeld[18] in the 1930's before its virtual extinction and the sharp change in survival rates.[45,92]

The females of the blue whale reach maturity at between four and seven years of age, and they have a gestation period of approximately one year. A single calf is born and is nursed for seven months during which period the female does not again become pregnant. This, together with the migratory habits of the species, implies that not more than one calf is born to a female every two years. The numbers of males and females are approximately equal and this ratio does not vary appreciably with age. The maximum age attained by a blue whale is assumed to be around 40 years. Survival rates, in the absence of exploitation, have been estimated as being about 87 per cent of the population each year for the first ten years of life.

If a two-year period is assumed, the Leslie matrix for the blue whale is as follows:

$$\mathbf{A} = \begin{bmatrix} 0 & 0 & 0{\cdot}19 & 0{\cdot}44 & 0{\cdot}50 & 0{\cdot}50 & 0{\cdot}45 \\ 0{\cdot}77 & 0 & 0 & 0 & 0 & 0 & 0 \\ 0 & 0{\cdot}77 & 0 & 0 & 0 & 0 & 0 \\ 0 & 0 & 0{\cdot}77 & 0 & 0 & 0 & 0 \\ 0 & 0 & 0 & 0{\cdot}77 & 0 & 0 & 0 \\ 0 & 0 & 0 & 0 & 0{\cdot}77 & 0 & 0 \\ 0 & 0 & 0 & 0 & 0 & 0{\cdot}77 & 0{\cdot}78 \end{bmatrix}$$

As before, the first row vector represents the average numbers of female offspring produced by each female in the two year periods, with the maximum fecundity occurring in the 8–9 year and 10–11 year age classes. Fecundity in the 12 year and older age class then declines slightly to an average of 0·45 per female. The natural mortality is assumed to be 13 per cent of the population in each two year period during the first ten years of life. The survival rate for whales in the 12 year and older age group is assumed to be 80 per cent.

The dominant eigenvalue and eigenvector of this matrix is again calculated by an iterative procedure as:

$$\lambda \text{ (for one year)} = \sqrt{1{\cdot}0072} = 1{\cdot}0036$$
$$\mathbf{a} = [1000,\ 764,\ 584,\ 447,\ 341,\ 261,\ 885]'$$

This eigenvalue indicates that the population is capable of increasing, and the natural logarithm of the eigenvalue is an estimate of the intrinsic rate of increase:

$$r = \ln \lambda = \ln 1{\cdot}0036 = 0{\cdot}0036$$

Similarly, the harvest that can be taken from this population can be estimated as:

$$H = 100 \left(\frac{\lambda - 1}{\lambda} \right) \text{ per cent} = 100 \left(\frac{1 \cdot 0072 - 1}{1 \cdot 0072} \right)$$

$$= 0 \cdot 71\% \text{ every two years or approximately } 0 \cdot 35\% \text{ per year.}$$

The eigenvalue being close to 1, the intrinsic rate of increase and, hence, the proportion of the population that can be harvested each year are small. If this harvest rate is exceeded, the species would inevitably decline unless homeostatic mechanisms were able to change the parameters of survival and fecundity.[62]

Usher[84] has investigated the effects of small changes in the survival and fecundity rates of the matrix. He shows that a decrease in all the survival rates of $0 \cdot 115$, from $0 \cdot 87$ to $0 \cdot 755$, results in the eingenvalue being reduced from $1 \cdot 0072$ to $1 \cdot 0000$. Similarly, a general decrease in fecundity of $48 \cdot 6$ per cent results in the eigenvalue being reduced from $1 \cdot 0072$ to $1 \cdot 0000$. In general, if the eigenvalue of a perturbed matrix is $\lambda + \delta\lambda$ and if the maximum element of $\delta\mathbf{A}$ does not exceed ε in absolute value, then

$$|\delta\lambda| < n\varepsilon/(\mathbf{y}'\mathbf{x})$$

gives limits on the effect of an error in $\mathbf{A}$ on λ, where $\mathbf{x}$ and $\mathbf{y}$ are normalized column eigenvectors of $\mathbf{A}$ and its transpose respectively. This expression often gives limits which are excessively wide, especially when many of the elements of the matrix are zero, as in this example. Nevertheless, the setting of the maximum effect of perturbations is a useful property, and further illustrates the advantage of accepting constraints on the formulation of the model.

Example 4.2 – Population structure of red deer.

As an example of the application of a matrix model to populations of both sexes, we will review various attempts to model the dynamics of red deer populations. The red deer (*Cervus elaphus*) is an animal that arrived in Britain during the Atlantic period and has played an important part in the general ecology as the largest of our wild ungulates. Originally, the population of these animals was controlled by predators of various kinds, including man. As man began to exert increasing control on the land of Britain, all of these predators were either controlled or killed out, being incompatible with agriculture, forestry and the safety of man himself. The progressive destruction of the forests which covered a large part of Scotland greatly changed the habitat available to the animals, and brought them into increasing competition with sheep and cattle on the grazing lands. During the nineteenth century, landowners maintained large herds of deer for deer-stalking at the expense of their own and their neighbours' agriculture and forestry. The increasing interest in upland forestry and agriculture, and a decreasing interest in deer-stalking have today placed considerable emphasis on the management of populations of red deer.

Lowe[51] has studied the dynamics of the population of red deer on the Island of Rhum since 1957. His estimate of the population structure of the deer on the island when he began his studies is summarized in Table 4.1. A life table has been constructed for this group of animals, for which the exploitation was very light and estimated as about 40 stags and 40 hinds each year, so that the population at this date may be regarded as under-exploited, the regulation of numbers being largely due to natural mortality. After the first year of life, the natural mortality increased to its highest point in the eight- and nine-year old animals, for both stags and hinds. There was also a marked difference between the mortalities of stags and hinds in the two to six year classes, the mortality rate for stags being only about one per cent while that for hinds was between 10 and 20 per cent.

Usher[83] has estimated the age-specific birth rates for the production of both male and female offspring, using Lowe's data on the proportion of the population breeding, and by taking weighted averages for the age groups. It is assumed that red deer produce only one calf each year.

Table 4.1 Population structure of red deer on the Island of Rhum in 1957. After Lowe.[51]

| | Numbers of animals | | Survival rates | |
	Stags	Hinds	Stags	Hinds
1	107·0	129·4	0·718	0·863
2	74·9	113·5	0·990	0·902
3	79·4	113·1	0·990	0·882
4	70·1	81·4	0·990	0·879
5	85·9	78·2	0·990	0·862
6	78·4	59·3	0·991	0·840
7	79·2	64·6	0·734	0·808
8	59·1	55·1	0·496	0·507
9	29·5	25·0	0·370	0·326
10	11·3	8·7	0·848	0·864
11	9·3	8·3	0·821	0·824
12	8·7	6·7	0·781	0·810
13	3·5	2·0	0·720	0·735
14	5·7	1·1	0·611	0·680
15	1·5	4·2	0·364	0·529
16	0·7	2·2	0·000	0·000
17	0·3	0		

Table 4.2 Matrix for both sexes of a red deer population.

$$\mathbf{A} =$$

Row 1: 0 0 0 0·202 0·419 0·434 0·362 0·363 0·355 0·376 0·422 0·417 0·464 0·464 0·464 0·464 0·464 0·464 0·464

Row 2: 0 0 0 0·214 0·444 0·459 0·589 0·589 0·576 0·612 0·353 0·348 0·388 0·388 0·388 0·388 0·388 0·388 0·388

Subdiagonal survival values: 0·718, 0·863, 0·990, 0·902, 0·990, 0·882, 0·879, 0·990, 0·990, 0·862, 0·991, 0·840, 0·734, 0·808, 0·496, 0·507, 0·370, 0·326, 0·848, 0·864, 0·821, 0·824, 0·781, 0·810, 0·720, 0·735, 0·611, 0·680, 0·364, 0·529

The matrix for both sexes of the red deer population is given in Table 4.2. The matrix is necessarily large, having 32 rows and columns. The dominant eigenvalue is as follows:

$$\lambda = 1 \cdot 1636$$

The corresponding eigenvector, representing the age structure of a stable population of red deer, and structured so that there are 1000 one-year stags, is summarized in Table 4.3. As

$$H = 100 \left(\frac{\lambda - 1}{\lambda} \right) \text{ per cent} = 100 \left(\frac{1 \cdot 1636 - 1}{1 \cdot 1636} \right) \text{ per cent} = 14 \cdot 1 \%$$

the model indicates that a harvest of 14.1 per cent can be taken each year from a virtually unexploited population, assuming that the same

Table 4.3 Eigenvector of red deer matrix.

Age (Years)	Stags	Hinds
1	1000	1239
2	617	919
3	525	712
4	447	540
5	380	408
6	323	302
7	275	218
8	174	151
9	74	66
10	24	18
11	17	14
12	12	10
13	8	7
14	5	4
15	3	2
16	1	1

percentage is taken from each age group. The model can equally well be used, however, to determine alternative strategies of harvesting in which the exploitation is concentrated on particular age classes.

Beddington[5] has investigated the exploitation of red deer in Scotland as a simple renewable resource, and as it affects the dynamics of the population. He found that the elements of the matrix **A** need to be considered as functions of the population density and of various environmental variables. Apart from weather variation, both survival and fecundity for each class were negatively correlated with increases in population density. He suggests, therefore, that the matrix model for red deer needs to be of the form:

$$\mathbf{n}_{t+1} = \mathbf{M}_{(N_t)}\mathbf{n}_t$$

where the subscript to the matrix $\mathbf{M}$ indicates that its elements are functions of N_t defined as:

$$N_t = \sum_i n_i$$

In the absence of exploitation, there is an equilibrium population density N^* given by the solution to:

$$\left|\mathbf{M}_{(N_t)} - \mathbf{I}\right| = 0$$

Extensive computer simulation of the population process also suggests that the neighbourhood stability analysis may describe a global stability, and that, in the absence of exploitation, the population would be expected to approach an equilibrium and fluctuations around this equilibrium due to variation would be damped.

By introducing a diagonal matrix $\mathbf{D}$, whose elements θ_i are the probabilities that the ith age class will survive the harvest, an extension of the Lefkovitch[48] equation is:

$$\mathbf{n}_{t+1} = \mathbf{M}_{(N_t)}\mathbf{D}\mathbf{n}_t$$

If the θ_i are held constant and are not so small that, at some level of the population, the population cannot increase, the population will reach an equilibrium defined by

$$\left|\mathbf{M}_{(N_t)}\mathbf{D} - \mathbf{I}\right| = 0$$

Example 4.3 – Size classes of Scots pine.

Forest trees are generally classified according to their size rather than their age. Usher[78] developed the basic Leslie model for selection forests, which contain an uneven age and size structure of the trees and in which there is no definite regeneration phase. In such forests, the tree will either remain in the same size class or move to the next largest size class, assuming that the time period is sufficiently small to ensure that the tree does not move up by more than one size class.

The major difficulty of the forest model is the analogue of the fecundity terms of the basic model. If we assume a forest regenerated naturally, any gap caused by the death or harvesting of a tree can be used either by natural regeneration or by the enlargement of the crowns of the surrounding trees. The fecundity terms of the matrix, therefore, depend not on the population of trees, but on the number of trees that are harvested. If there are n_i trees in size class i at time t, then at $t+1$, when we have a stable population structure, there should be λn_i trees in that class, and $(\lambda-1)n_i$ of these trees will be harvested to reduce the population to n_i. The factor $(\lambda - 1)$ is therefore required to modify the regeneration terms in the first row of the matrix model.

Usher's[80] model is:

$$Q = \begin{bmatrix} a_0 + c_0(\lambda-1) & c_1(\lambda-1) & c_2(\lambda-1) & \ldots & c_{n-1}(\lambda-1) & c_n(\lambda-a_n) \\ b_0 & a_1 & 0 & \ldots & 0 & 0 \\ 0 & b_1 & a_2 & \ldots & 0 & 0 \\ 0 & \cdot & \cdot & & \cdot & \cdot \\ \cdot & \cdot & \cdot & & \cdot & \cdot \\ \cdot & \cdot & \cdot & & \cdot & \cdot \\ 0 & 0 & 0 & \ldots & a_{n-1} & 0 \\ 0 & 0 & 0 & \ldots & b_{n-1} & a_n \end{bmatrix}$$

where a_i $(i=0, 1, \ldots, n-1)$ is the probability that a tree will remain in the ith size class; b_i $(i=0, 1, \ldots, n-1)$ is the probability that a tree will move up one size class; c_i $(i=0, 1, \ldots, n)$ is the number of trees of size class o regenerating in the gap caused by the harvesting of a tree. a_n is a management decision: it depends on the number of trees to be left in the largest class and will often be o. This in turn affects the regeneration term which becomes $c_n(\lambda - 1 + b_n)$ if $b_n = 1 - a_n$, and thus this simplifies to $c_n(\lambda - a_n)$.

The calculation of the primary eigenvalue and eigenvector of a matrix which contains functions of its own latent root requires a more complex method than the iterative procedure which we have used so far. Usher[80,81] gives a simple and reasonably fast method of solution and shows that the model has only one eigenvalue greater than unity and an associated eigenvector whose elements are all non-negative. There is, therefore, only one size structure for the forest that is biologically meaningful and maximizes the production of the forest.

Usher[78] also gives an example of the application of this model to a Scots pine forest, based on data derived from the forest and from calculations of the ground area occupied by trees of various sizes.[37] The basic matrix derived is as follows:

$$A = \begin{bmatrix} 0.72 & 0 & 0 & 3.6(\lambda-1) & 5.1(\lambda-1) & 7.5\lambda \\ 0.28 & 0.69 & 0 & 0 & 0 & 0 \\ 0 & 0.31 & 0.75 & 0 & 0 & 0 \\ 0 & 0 & 0.25 & 0.77 & 0 & 0 \\ 0 & 0 & 0 & 0.23 & 0.63 & 0 \\ 0 & 0 & 0 & 0 & 0.37 & 0 \end{bmatrix}$$

The eigenvalue and eigenvector of this matrix is calculated as

$$\lambda = 1\cdot204266$$
$$\mathbf{a} = [1000,\ 544,\ 372,\ 214,\ 86,\ 26]'$$

The matrix assumes time periods of six years, and as

$$H = 100 \left(\frac{1\cdot204266 - 1}{1\cdot204266} \right) \text{ per cent} = 16\cdot96\%$$

harvesting of approximately 17 per cent of the trees can be carried out in every six-year period.

Usher[84] has extended this particular model from six to eight size classes and has used the model to investigate harvesting strategies and, particularly, the maximum size to which a tree should be allowed to grow. His provisional conclusion was that short rotations produce a smaller annual increment than longer rotations.

Example 4.4 – Phosphorus cycle in a three-compartment eco-system.

In this final example in this chapter on matrix models, we will consider the use of such models to simulate energy flow and nutrient cycles in ecosystems. These are clearly dynamic processes of the ecosystems, and some modification of our basic matrix is necessary to represent the input of energy or nutrients into the ecosystem and the transfer of energy or nutrients within the ecosystem. It is also clearly convenient to assume some arbitrary compartmentation of the ecosystem into the species or trophic levels of the system, or into the functional parts of these species. Finally, we do not need to specify losses from the ecosystem in our model; we can assume that these losses are the differences between the inputs and the sums of the outputs from, and storage within, the specified compartments.

Both Usher[78] and Goodman[28] have commented on the fact that the basic Leslie matrix may be considered as the sum of two matrices:

$$
\mathbf{A} =
\begin{bmatrix}
f_0 & f_1 & f_2 & \cdots & f_n \\
0 & 0 & 0 & \cdots & 0 \\
0 & 0 & 0 & \cdots & 0 \\
\cdot & \cdot & \cdot & & \cdot \\
\cdot & \cdot & \cdot & & \cdot \\
\cdot & \cdot & \cdot & & \cdot \\
0 & 0 & 0 & \cdots & 0
\end{bmatrix}
+
\begin{bmatrix}
0 & 0 & 0 & \cdots & 0 \\
p_0 & 0 & 0 & \cdots & 0 \\
0 & p_1 & 0 & \cdots & 0 \\
\cdot & \cdot & \cdot & & \cdot \\
\cdot & \cdot & \cdot & & \cdot \\
\cdot & \cdot & \cdot & & \cdot \\
0 & 0 & 0 & \cdots & 0
\end{bmatrix}
$$

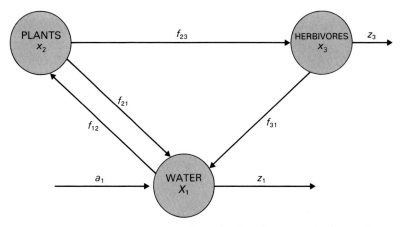

Fig. 4.2   Diagrammatic representation of phosphorus cycle in a three-compartment ecosystem.

or

$$A = F + P$$

where the matrix F represents the input of new members to the population, and the matrix P represents the transition of members between the age classes of the population. We can exploit the partition of the basic matrix in our representation of dynamic processes.

Smith[73] developed a model for the flow of phosphorus in a three-compartment system, and a diagram of this system is given in Fig. 4.2. The model is defined by five categories of parameters:

x_i = the amount of phosphorus in the ith compartment at any specified time;

a_i = the rate of inflow of phosphorus into the ith compartment;

z_i = the rate of outflow of phosphorus from the ith compartment;

f_{ij} = the rate of flow of phosphorus from the ith to the jth compartment;

f_{ii} = the proportion of phosphorus that is stored in the ith compartment during a period of time.

The matrix model is:

$$\begin{bmatrix} f_{11}+a_1/x_{t,1} & f_{21} & f_{31} \\ f_{12} & f_{22}+a_2/x_{t,2} & f_{32} \\ f_{13} & f_{23} & f_{33}+a_3/x_{t,3} \end{bmatrix} \begin{bmatrix} x_{t,1} \\ x_{t,2} \\ x_{t,3} \end{bmatrix} = \begin{bmatrix} x_{t+1,1} \\ x_{t+1,2} \\ x_{t+1,3} \end{bmatrix}$$

where the two vectors denote the amounts of energy or nutrients in the ith compartment at times t and $t+1$. The matrix elements not on the

Table 4.4   Parameters of phosphorus cycle model.

Parameter	Function
x_1	9·5
x_2	1·4
x_3	9·0
f_{12}	$0·10417x_1x_2$
f_{21}	$0·05208x_2$
f_{23}	$0·10417x_2x_3$
f_{31}	$0·05208x_3$
z_1	$0·02083x_1$
z_3	$0·01042x_3^2$
f_{11}	$(1-0·02083-0·10417x_2)x_1$
f_{22}	$(1-0·05208-0·10417x_3)x_2$
f_{33}	$(1-0·05208-0·01042x_3)x_3$

principal diagonal give the transitions between the ith and jth compartments. The elements of the leading diagonal are made up from
two sources: the energy or nutrients not transferring between compartments, and the input into the ith compartment independent of the
amount of energy or nutrients already in that compartment, and hence
represented by $a_i/x_{t,1}$.

Usher[82] has derived the parameters of this model, converted to a
quarter-hour time period, from the data originally published by
Smith,[73] and these parameters are given in Table 4.4. The rates of
flow are converted to flows per unit of nutrient by dividing them by the
appropriate total nutrient in the ith compartment, i.e. by x_i. The basic
matrix then becomes:

$$\mathbf{A} = \begin{bmatrix} 0·94298 & 0·05208 & 0·05208 \\ 0·14584 & 0·01042 & 0 \\ 0 & 0·93753 & 0·85417 \end{bmatrix}$$

The dominant eigenvalue of this matrix is 1, with an associated eigenvector of $[9·5, 1·4, 9·0]'$. In other words, the analysis confirms that the
ecosystem is in a steady state, with no increase in the total amount of
phosphate.

It should perhaps be stressed that matrix models of the flow of
energy or nutrients may occasionally lead to problems in the calculation
of the dominant eigenvalue and eigenvector, particularly when the
matrix elements become negative. These difficulties can usually be
overcome by a careful choice of the time period over which the matrix
operates. Usher[82] gives some additional examples.

In this chapter on matrix models, we have explored one family of models in which the 'realism' of the model is partly sacrificed in order to obtain the benefit of the mathematical formulation. The same formulation also imposes constraints upon the way in which the models can be used, but these constraints are balanced by the convenience of the computations and by the relative ease of establishing the values of the basic parameters.

Matrix, or linear, algebra is one of the most powerful operational tools of modern mathematics. Basically, it has four advantages:

1. It summarizes expressions and equations very compactly.
2. It facilitates the memorizing of these expressions.
3. It greatly simplifies the procedures for deriving solutions to complex problems.
4. High-level computer languages have convenient instructions for the handling of matrix computation.

Although matrix calculations are sometimes extensive, especially in matrix inversion and in the calculation of eigenvalues and eigenvectors, and will often require the use of computers, these calculations are usually very much less difficult to program than those involved in dynamic models. Furthermore, the properties of the basic matrices of the models enable the modeller to exploit the deductive logic of pure mathematics. In this chapter, we have confined our attention to the dominant eigenvalue and eigenvector as expressions of the basic matrix. The remaining eigenvalues and eigenvectors may be used as expressions of the stability or oscillatory tendency of the models, but have been omitted from this introduction because of the relative complexity of the computer program to calculate all of the eigenvalues and eigenvectors of the matrix.

Matrix models represent an important, and neglected, family of models in systems analysis. So far, this family has been relatively unexploited in the ecological sciences, and only a few research workers have published applications of such models. In part, the unfamiliarity of biologists and ecologists with matrix algebra has been responsible for the neglect of matrix models, although the differential and difference equations of dynamic models make an even greater demand on the mathematical ability of the modeller. The increasing availability of computers, and the realization that even small computers (i.e. the mini- and micro-computers of the present-day computer revolution) can be programmed to calculate at least the dominant eigenvalue and eigenvector will probably result in a wider application of matrix formulations.

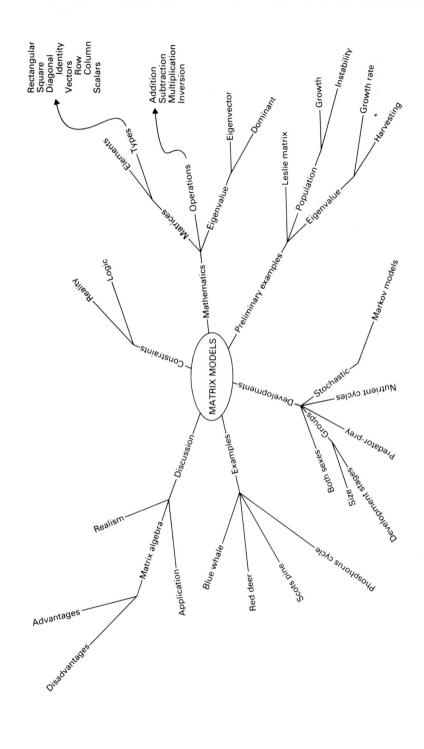

5

Stochastic Models

The families of models which we have so far considered have all been *deterministic*. That is to say that, from a given starting point, the outcome of the modelled response is necessarily the same and is predicted by the mathematical relationships incorporated in the model. Deterministic models are the logical development of the kinds of mathematics we learn early in our mathematics courses, and particularly in what has come to be called 'applied mathematics', i.e. mathematics applied to physics. Such models are necessarily mathematical analogues of physical processes in which there is a one-to-one correspondence between cause and effect.

There is, however, a later development of mathematics which enables relationships to be expressed in terms of probabilities, and in which the outcome of a modelled response is not certain. Models which incorporate probabilities are known as *stochastic* models, and such models are particularly valuable in simulating the variability and complexity of ecological systems. Probabilities can, of course, be introduced into almost any kind of model, for example in the dynamic models of Chapter 3, and particularly in the study of the stability of such models to variations in the basic parameters. In this chapter, however, we will confine our attention to models which are primarily stochastic. As the development of statistical theory has a long history, we will necessarily have to be selective in the number of types presented.

SPATIAL PATTERNS OF ORGANISMS

One of the simplest applications of stochastic models to ecological problems is the mapping of spatial patterns of living organisms. There is, of course, a wide range of statistical distributions capable of describing spatial patterns and we will consider only the simplest of these distributions in this text.

The problem arises from the frequent need in ecology to understand and to predict the numbers of organisms which will be found on some defined area, or on some equally well-defined unit, for example a single plant, a single leaf or a single seed. In general, we will not know the average density of the organisms, and we therefore need a mathematical model which will provide us with an efficient measure of the average number of occurrences and which will also provide us with a measure of the variability of the occurrences and the pattern of this variability.

If we ignore for a moment the practical possibility of a totally uniform distribution, i.e. that each sampling unit has exactly the same number of individuals, with no variation, the simplest hypothesis that we can form is that each organism has a constant, but unknown, probability of occurring in the unit and that the presence of the individuals has no effect upon their neighbours, or, in other words, that the probabilities of occurrence of individuals are independent. Without working through the derivation, which can, however, be found in most texts on elementary probability theory (e.g. Weatherburn[93]) we can show that, under these assumptions, the probabilities of o, 1, 2, 3, . . ., x individuals per sampling unit are given by the series

$$e^{-m}, \ e^{-m}\frac{m}{1!}, \ e^{-m}\frac{m^2}{2!}, \ e^{-m}\frac{m^3}{3!}, \ \ldots, \ e^{-m}\frac{m^x}{x!},$$

where e is the base of natural (Naperian) logarithms.

This series is well-known in the statistical literature as the Poisson Series or Poisson distribution. The average number of individuals per sampling unit from this series is m and the variance of the number of individuals is also m. Tests of the adequacy of the Poisson distribution to describe the spatial distribution of the organisms (and hence of the hypothesis that the probability of occurrence of an individual is constant and not affected by the presence of other individuals) can be made either by comparing the observed frequency of occurrences with the frequency that would be expected from the theoretical distribution or by comparing the mean and the variance of the observed frequencies.

If the hypothesis of the Poisson distribution is rejected, we may then be justified in formulating some alternative hypothesis about the distribution. While this formulation may be guided by what we know about the organisms, there is always an infinity of possible hypotheses, and our search for an adequate mathematical model should therefore be consistent with the ecology of the problem rather than with the convenience of the mathematics. Similarly, the model should not be over-elaborate, nor require more parameters than can be estimated from any reasonable set of data. Many of the alternatives to the Poisson distribution provide models for quite specific departures from randomness, and their appropriateness therefore depends upon the underlying ecology.

If the individual organisms move away from each other, especially as the number of individuals increases, the characteristic feature of the distribution will be its regularity and the uniform spacing of the individuals, with the variance of the number of individuals becoming smaller than the mean number of individuals. Territorial behaviour of animals, for example, will often produce a relatively uniform spacing of individuals and the dispersion of sedentary invertebrates may be regular over a small area of stream or lake bottom. A regular distribution of this kind may be approximated by the positive binomial distribution for which the expected frequency distribution is given by the expansion of:

$$n(q+p)^k$$

where n = number of sampling units

       p = probability of any point in the sampling unit being occupied by an individual

       $q = 1 - p$, and

       k = maximum possible number of individuals a sampling unit could contain.

In practice, estimates of the parameters k, p and q are obtained from samples of the population. A test of the conformity of observed counts with the positive binomial distribution is again provided by a comparison of the observed and expected frequencies, using a χ^2 test of goodness of fit.

Where the spatial distribution of individuals is neither random nor regular, and the variance of the numbers of individuals per sample unit is greater than the mean number of individuals per unit, the distribution is usually referred to as 'contagious', indicating that there are clumps or patches of individuals and irregular gaps with no individuals. There are, of course, many environmental factors which can contribute to uneven distributions of individuals, and there is frequently a tendency for some species to aggregate and thus produce clumping even without the influence of environmental factors. The resulting pattern of distribution is dependent upon the size of the groups, the distance between the groups, the spatial distribution of the groups and the spatial distribution of individuals within groups.

Several mathematical models have been suggested for irregular distributions, and, of these, the negative binomial is perhaps the best known. This distribution is given by the expansion of:

$$(q-p)^{-k}$$

where $p = \mu/k$
$q = 1 + p$

The parameters of this distribution are the arithmetic mean μ and the exponent k. This exponent related to the spatial distribution of the individuals and its inverse $1/k$ is a measure of the excess variance or grouping of the individuals in the population. As $1/k$ approaches zero and k approaches infinity, the distribution converges to the Poisson series and to random occurrence of the individuals. If, however, $1/k$ approaches infinity and k approaches zero, the distribution converges to the logarithmic series whose properties and applications have been described by Williams.[94]

Various tests are available for fitting the negative binomial distribution, the efficiency of the various tests depending upon the size of the sample, the average number of individuals per sample and the ease of estimating k. The selection of appropriate methods is too complex a diversion for this text and is discussed in some detail by Elliott.[19]

The disadvantage of the negative binomial distribution is that it can be derived from a wide variety of biological models. It may, for example, arise from a model in which there is true contagion, in the sense that the presence of one individual in a sampling unit increases the chance that another individual will also occur. Similarly, it can be shown that the growth of a population with constant birth and death rates of individuals and a constant rate of immigration leads to a negative binomial distribution of population size. Groups of individuals distributed at random with the numbers of individuals in the groups distributed in a logarithmic distribution and populations made up of several sub-populations, each distributed at random but with different probabilities of occurrence, also follow a negative binomial distribution.

Although the negative binomial may be successfully fitted to a wide variety of distributions, the inability to identify the distribution with any one set of initial circumstances makes it only partially useful—an example of a model whose adaptability is a disadvantage! For this reason, a number of other contagious distributions have been developed. For example, Anscombe[1] suggests the Thomas,[77] Neyman[58] Type A, Polya,[63] and the discrete log-normal distributions as possible alternatives for distributions of organisms whose variance is greater than the mean and for which the asymmetry is positively skewed.

The Thomas and Neyman distributions are very similar, but are derived from different assumptions. Neyman's distribution is intended to describe the dispersion of organisms which have originated from randomly distributed clumps, the arbitrary limit to the clump size being determined by the distance through which the organisms have dispersed. Thomas' distribution is essentially a double Poisson distribution, one distribution describing the number of clumps and the other numbers of individuals in the clumps. Both of these models have the capacity to describe distributions with more than one mode, whereas the negative binomial distribution always has only one mode.

The Polya distribution is derived from the simultaneous and random colonization of some particular habitat by parent organisms. These parent organisms produce clusters of offspring, the numbers of individuals in a cluster following a geometric distribution. The discrete log-normal distribution is simply derived by a straight logarithmic transformation of the counts of the numbers of organisms in each sample unit.

As the number of organisms per sample unit increases, or, alternatively, as the number of sample units increases, all of these distributions approach closer and closer to the Normal distribution which is fundamental to many of the basic ideas and tests of statistical mathematics. Even for quite small numbers of samples, approximation to the Normal distribution can be achieved by relatively simple transformations of the observed counts, and the identification of the correct transformation may itself provide an adequate guide to the nature of the distribution of the organisms. It should also be mentioned that alternative methods of investigating spatial distributions have been developed from measurements of the distances between neighbouring individuals. These methods are outside the scope of this introductory text, but Pielou[61] and Greig-Smith[30] summarize some of the earlier techniques.

Example 5.1 – Numbers of *Helobdella* in samples from a lake.

Table 5.1 gives the numbers of the leech *Helobdella* found in 103 samples taken from a freshwater lake, and the frequencies of these samples are summarized in Table 5.2.

The total number of the organisms found was 84, so that the mean number per sample may be calculated as:

$$\bar{x} = \frac{84}{103} = 0.8155$$

Table 5.1   Numbers of *Helobdella* in 103 samples from a freshwater lake.

0	0	1	0	0	1	3	0	0	1
3	2	0	2	2	2	0	0	0	2
0	1	0	0	0	0	0	1	0	1
1	1	0	0	0	0	6	2	2	0
1	0	2	0	1	0	1	2	1	0
0	1	4	0	0	0	1	0	4	0
2	1	1	1	5	0	0	0	0	0
0	0	0	0	2	1	1	0	0	2
0	0	0	1	0	2	1	0	0	1
0	1	8	0	1	0	0	0	0	0
0	1	0	—	—	—	—	—	—	—

Table 5.2   Comparison of observed frequencies of *Helobdella* with frequencies expected from a Poisson distribution.

Number of organisms	Observed frequency	Expected frequency	Differences
0	58	45·57	−12·43
1	25	37·16	+12·16
2	13	15·15	+2·15
3	2	4·12	+2·12
4	2	0·84 ⎫	
5	1	0·14 ⎪	
6	1	0·02 ⎬ 1	−4
7	0	0·00 ⎪	
8	1	0·00 ⎭	

Under the hypothesis that the probability of the occurrence of an individual remains constant and independent, the frequency of occurrence of 0, 1, 2, 3, etc. organisms is calculated from the successive terms of the Poisson series.

$$P_{(x)} = e^{-m} \frac{m^x}{x!}$$

where $m = \bar{x}$.

The expected frequency of 0 individuals per sampling unit is:

$$n\,e^{-m} = 103(0\cdot442418) = 45\cdot57$$

and the remaining frequencies from the successive terms are as follows:

$$n\,e^{-m}\,m = 103(0.442418)(0.8155) = 37.16$$

$$n\,e^{-m}\,\frac{m^2}{2!} = 103(0.442418)(0.3325) = 15.15$$

$$n\,e^{-m}\,\frac{m^3}{3!} = 103(0.442418)(0.0904) = 4.12$$

$$n\,e^{-m}\,\frac{m^4}{4!} = 103(0.442418)(0.0184) = 0.84$$

$$n\,e^{-m}\,\frac{m^5}{5!} = 103(0.442418)(0.0030) = 0.14$$

$$n\,e^{-m}\,\frac{m^6}{6!} = 103(0.442418)(0.0004) = 0.02$$

These frequencies are summarized in Table 5.2, together with the deviations of the observed from expected frequencies. The numbers of organisms with small expected frequencies have been pooled to give a minimum expected frequency of 1. The test of the adequacy of the Poisson distribution as a model for the distribution of *Helobdella* in the lake samples is then made by the χ^2 test described in any textbook on elementary statistics (e.g. Balaam[4]). The test is calculated as

$$\chi^2 = \sum_{i=1}^{k} \frac{(O_i - E_i)^2}{E_i}$$

where k is the number of classes
       O_i is the observed frequency in class i
       E_i is the expected frequency in class i

For the data of our example,

$$\chi^2 = \frac{(-12.43)^2}{45.57} + \frac{(12.16)^2}{37.16} + \frac{(2.15)^2}{15.15} + \frac{(2.12)^2}{4.12} + \frac{(-4)^2}{1} = 24.77$$

This value of χ^2, compared with the tabulated values for $(k-1) = 3$ degrees of freedom, suggests that the probability of obtaining a sample which deviated so far from the expected values for a population of individuals with a constant probability of occurring is less than 0.001 (actually close to 0.0003), and would usually be regarded as negligible.

An alternative test is provided by the comparison of the variance of the observed numbers of organisms with the mean number of individuals per sample.

$$\chi^2 = \frac{s^2(n-1)}{m}$$

where s^2 is the variance
   n is the number of samples
   m is the mean number of organisms per sample.

Again, for example

$$\chi^2 = \frac{1.7990(103-1)}{0.8155} = 225.01$$

with 102 degrees of freedom.

The probability of achieving a value of χ^2 as high as this is again very much less than 0.001, and the model of a constant probability of an individual occurring must, therefore, be rejected.

The reader should note that, in the last few paragraphs, two different uses of probabilities have been introduced, a fact which may lead to some confusion. In the first place, we have defined a stochastic model in which there is a constant probability of an individual occurring resulting in an average of 0.8155 individuals occurring per sample. In order to test the validity of the model, we have used a statistical test of the goodness of fit of the observed frequencies to those expected by the model. The results of this test are expressed as a probability (not to be confused with the probability of the occurrence of an individual organism) that the sample data could have been obtained from a population with the properties of the model. As a convention, we usually reject any hypothetical model which has a probability of less than 0.05 of accounting for the observed values.

In order to fit a negative binomial distribution to the data of this example, we first need an estimate of k in the expression

$$P_{(x)} = \left(1 - \frac{\mu}{k}\right)^{-k} \frac{(k+x-1)!}{x!(k-1)!} \cdots \left(\frac{\mu}{\mu+k}\right)^x$$

where $P_{(x)}$ is the probability of x individuals in a sampling unit and the parameters of μ and k are estimated from the frequency distribution of the sample by the statistics $\bar{x}$ and k. A preliminary estimate of k may be derived from the equation:

$$k = \frac{\bar{x}^2}{s^2 - \bar{x}} = \frac{(0.8155)^2}{(1.7990 - 0.8155)} = 0.68$$

but this estimate should then be used as the starting point for substitution in the maximum-likelihood equation

$$n \log_e \left(1 + \frac{\bar{x}}{k} \right) = \Sigma \left(\frac{A_{(x)}}{k + x} \right)$$

where n is the total number of sampling units, $\log_e$ denotes the natural (Naperian) logarithm, and $A_{(x)}$ is the total number of counts exceeding x. Different values of k are tried until the equation is approximately balanced, and, for the data of this example, $k = 0.8$ is a reasonable solution.

The individual terms of the frequency distribution are calculated progressively by the stages given below:

x	$P_{(x)}$		$f = n P_{(x)}$
0	$P_{(x=0)} = \left(1 + \dfrac{\bar{x}}{k} \right)^{-k}$	$= 0.5699$	58.70
1	$P_{(x=1)} = \left(\dfrac{k}{1} \right) \left(\dfrac{\bar{x}}{x+k} \right) P_{(x=0)}$	$= 0.2302$	23.71
2	$P_{(x=2)} = \left(\dfrac{k+1}{2} \right) \left(\dfrac{\bar{x}}{x+k} \right) P_{(x=1)} = 0.1046$		10.77
3	$P_{(x=3)} = \left(\dfrac{k+2}{3} \right) \left(\dfrac{\bar{x}}{x+k} \right) P_{(x=2)} = 0.0493$		5.07
4	$P_{(x=4)} = \left(\dfrac{k+3}{4} \right) \left(\dfrac{\bar{x}}{x+k} \right) P_{(x=3)} = 0.0236$		2.43
5	$P_{(x=5)} = \left(\dfrac{k+4}{5} \right) \left(\dfrac{\bar{x}}{x+k} \right) P_{(x=4)} = 0.0114$		1.18
6	$P_{(x=6)} = \left(\dfrac{k+5}{6} \right) \left(\dfrac{\bar{x}}{x+k} \right) P_{(x=5)} = 0.0056$		0.58
7	$P_{(x=7)} = \left(\dfrac{k+6}{7} \right) \left(\dfrac{\bar{x}}{x+k} \right) P_{(x=6)} = 0.0027$		0.28
8	$P_{(x=8)} = \left(\dfrac{k+7}{8} \right) \left(\dfrac{\bar{x}}{x+k} \right) P_{(x=7)} = 0.0013$		0.14

The resulting expected frequencies are summarized in Table 5.3, together with the observed frequencies and the differences between observed and expected frequencies. The results scarcely need any

Table 5.3 Comparison of observed frequencies with frequencies predicted from negative binomial distribution.

Number of organisms	Observed frequency	Expected frequency	Differences
0	58	58·70	−0·70
1	25	23·71	1·29
2	13	10·77	2·23
3	2	5·07	−3·07
4	2	2·43	−0·43
5	1	1·18 ⎫	
6	1	0·58 ⎬ 2·32	0·68
7	0	0·28 ⎪	
8	1	0·14 ⎭	
Totals	103	103	0·00

confirmation of the greatly improved fit of this alternative model but the chi-square test is calculated as:

$$\chi^2 = \frac{(-0\cdot70)^2}{58\cdot70} + \frac{(1\cdot29)^2}{23\cdot71} + \frac{(2\cdot23)^2}{10\cdot77} + \frac{(-3\cdot07)^2}{5\cdot07}$$

$$+ \frac{(-0\cdot43)^2}{2\cdot43} + \cdots + \frac{(0\cdot68)^2}{2\cdot32} = 2\cdot67$$

with four degrees of freedom (i.e. $n-2$, because the model now has two parameters, μ and k). The probability of a value of χ^2 as extreme as this for a sample drawn from a population following the negative binomial distribution is approximately 0·70, and certainly not sufficiently small for the hypothesis to be rejected.

As a result of these calculations, which appear more extensive than they really are because we have given each step in full, we may say that the expansion of the negative binomial

$$(q-p)^{-k}$$

provides an adequate model of the distribution of the numbers of *Helobdella* in the samples from this particular freshwater lake, where

$$k = 0\cdot8$$
$$p = \bar{x}/k = 1\cdot0194$$
$$q = 1 + p$$

The individual terms of the model are given by:

$$P_{(x)} = \left(1 + \frac{\bar{x}}{k}\right)^{-k} \frac{(k+x-1)!}{x!(k-1)!} \left(\frac{\bar{x}}{x+k}\right)^x$$

where $P_{(x)}$ is the probability of x individuals in the sampling unit, and, again, where

$$k = 0{\cdot}8$$
$$\bar{x} = 0{\cdot}8155$$

ANALYSIS OF VARIANCE

One of the most widely used stochastic models in scientific research is the one which underlies the statistical technique of analysis of variance, although many scientists who use statistical methods are scarcely aware that they are using a model, perhaps because this aspect of the analysis is seldom emphasized in elementary statistical texts. Nevertheless, the linear and factorial models underlying the analysis of variance have been of considerable importance in the development of science, and will probably remain important, despite the limitations of their basic assumptions. Again, the development of these models in the design and analysis of experiments and surveys has been one of the major achievements of the last fifty years, and we can, therefore, examine only a simple example of their application.

The basic model assumes a limited number of independent factors or possible effects which, added to the average, are capable of defining the practical situation to be modelled. Thus, a simple experiment with t treatments repeated in r separate blocks might be defined by the model

$$Y_{ij} = \mu + \beta_i + \tau_j + \varepsilon_{ij}$$

where μ is the mean
         β is the effect of the ith block, where $i = 1$ to r
         τ is the effect of the jth treatment, where $j = 1$ to t
and ε are random errors which are normally and independently distributed with mean zero and variance σ^2.

This model can be considerably simplified by further assuming that

$$\beta_1 + \beta_2 + \beta_3 + \cdots + \beta_r = 0$$

and

$$\tau_1 + \tau_2 + \tau_3 + \cdots + \tau_t = 0$$

In general, of course, we do not know the values of the various parameters of the model, and these have to be derived from samples of some defined population in such a way that our estimates are unbiased values for the population. One of the conditions for deriving unbiased estimates of this kind is that there must be an element of randomness in the design of the experiment; in the model described above, for example, the t treatments must be allocated to the individual plots (or other units of experimental material) at random within each block. Subject to this constraint, it is customary to derive the estimates of the population parameters from the criterion of least squares, i.e. to use those values of the model parameters μ, β_i, τ_j which give the smallest sum of squares of deviations from the observed values. The theory is relatively complex, and may be found in any statistical text, for example Balaam.[4] A simple example will help to show the application to practical problems.

Example 5.2 – Growth of seedlings of Sitka spruce on sterilized and unsterilized soil.

In many ecological situations, the growth of first year seedlings is inhibited by the activities of other organisms, even where competition from established vegetation is absent. Even in carefully prepared seedbeds, for example, the growth of one-year seedlings of Sitka spruce, *Picea sitchensis* (Bong.) Carr, in both height and diameter may be disappointingly small, and suggests the possibility of measuring the effect of other organisms by comparing the growth of seedlings on seedbeds prepared normally with that on seedbeds on which the soil has previously been sterilized. As a slight complication, let us further assume that we wish to test two alternative methods of sterilization, i.e. sterilization by steam and sterilization by the chemical substance formalin. We now have three treatments, viz:

$$T_1 = \text{control, unsterilized}$$
$$T_2 = \text{sterilized by formalin}$$
$$T_3 = \text{sterilized by steam.}$$

In order to determine the values of the parameters for some particular soil population, we will need to compare these three treatments, carefully randomized, on at least two separate occasions, or on at least two separate samples of soil from the same population. (The conditions for valid experiments are actually more complex than this, but are outside the scope of this text—see, for example, Fisher,[22] Federer,[20] or Pearce.[60]) In this example, we will repeat the experiment on eight separate samples of soil, each of which is subsequently divided into three parts to be allocated at random to the experimental treatments. Let us further assume that, at the end of the year, the mean diameters of the seedlings grown on each plot are as summarized in Table 5.4.

The mean for the whole experiment is estimated as the mean of the sample represented by Table 5.4, derived by adding together the 24

Table 5.4 Mean diameters of one-year seedlings of Sitka spruce grown on sterilized and unsterilized soil.

| Block | Mean diameter (millimetres) | | | Block totals |
	Control	Formalin	Steam	
1	0·84	1·09	1·16	3·09
2	0·84	1·03	1·66	3·53
3	0·97	0·88	1·50	3·35
4	1·06	0·94	1·50	3·50
5	1·00	1·13	1·59	3·72
6	0·81	1·00	1·31	3·12
7	0·75	1·06	1·44	3·25
8	0·91	0·56	1·50	2·97
Treatment totals	7·18	7·69	11·66	26·53

values and dividing by 24, i.e.

$$m = 26{\cdot}53/24 = 1{\cdot}105 \text{ mm}$$

Similarly, the effects of blocks 1, 2, 3, ..., 8 may be estimated from the means of the blocks, so that the effect of block 1 is estimated as:

$$B_1 = (3{\cdot}09/3) - 1{\cdot}105 = 1{\cdot}030 - 1{\cdot}105 = -0{\cdot}075 \text{ mm}$$

Remember that we simplified the model by making the assumption that the sum of the block effects should be zero, so that we have to subtract the experimental means from each block effect. Similarly, the effect of the second block is estimated as:

$$B_2 = (3{\cdot}53/3) - 1{\cdot}105 = 1{\cdot}177 - 1{\cdot}105 = 0{\cdot}072 \text{ mm}$$

and so on for the remaining blocks.

The effects of the three treatments are calculated in a similar way. Thus,

$$T_1 = (7{\cdot}18/8) \ - 1{\cdot}105 = 0{\cdot}898 - 1{\cdot}105 = -0{\cdot}208$$
$$T_2 = (7{\cdot}69/8) \ - 1{\cdot}105 = 0{\cdot}961 - 1{\cdot}105 = -0{\cdot}144$$
$$T_3 = (11{\cdot}66/8) - 1{\cdot}105 = 1{\cdot}458 - 1{\cdot}105 = 0{\cdot}353$$

and

$$T_1 + T_2 + T_3 = (-0{\cdot}208) + (-0{\cdot}144) + (0{\cdot}353) = 0{\cdot}001$$

i.e. nearly zero.

(Note that we can derive the estimates of the block and treatment effects in this simple way only when the experimental design has the property of orthogonality. The ease of the derivation is, indeed, one of the reasons for the importance of orthogonal experimental designs, particularly before electronic computers became widely available and more complex computational procedures therefore became possible. Orthogonal designs are, however, still important even with the ready availability of computers, as they greatly increase the ease of analysis of complex models and experiments.)

We can now calculate the expected value for each plot of the experiment and find the deviations between the observed and expected values. For the three treatments of the first block, for example, the expected values are:

$$\hat{Y}_{11} = m + B_1 + T_1$$
$$\hat{Y}_{12} = m + B_1 + T_2$$
$$\hat{Y}_{13} = m + B_1 + T_3$$

or

$$\hat{Y}_{11} = 1 \cdot 105 + (-0 \cdot 075) + (-0 \cdot 208) = 0 \cdot 822 \text{ mm}$$
$$\hat{Y}_{12} = 1 \cdot 105 + (-0 \cdot 075) + (-0 \cdot 144) = 0 \cdot 886 \text{ mm}$$
$$\hat{Y}_{13} = 1 \cdot 105 + (-0 \cdot 075) + (0 \cdot 353) = 1 \cdot 383 \text{ mm}$$

and the deviations of observed from expected values are:

$$d_{11} = 0 \cdot 84 - 0 \cdot 822 = 0 \cdot 018 \text{ mm}$$
$$d_{12} = 1 \cdot 09 - 0 \cdot 886 = 0 \cdot 204 \text{ mm}$$
$$d_{13} = 1 \cdot 16 - 1 \cdot 383 = -0 \cdot 223 \text{ mm}$$

Tables 5.5 and 5.6 give BASIC programs for calculating the additive effects of a wide range of designs and for recombining these effects to give the predicted values. These two processes are so fundamental to the analysis of additive models that they form the basis for a wide range of procedures and can even be used for estimating missing values in otherwise orthogonal experimental designs. Table 5.7 shows the results given by these programs for the data of this example.

The main value of the additive model described above, however, lies in the extension of the technique to the analysis of the variance defined by the model. How many of the parameters are really necessary in describing the response of the seedlings to the three treatments— assuming for the moment that the application of no sterilization—the control—is a form of treatment? In most practical situations, we are not really interested in the block effects, as the blocks were introduced to give us the necessary replication of the treatment comparisons, and may, in addition, have been used to increase the precision of the comparisons. For example, some of the blocks may contain soil collected on a different date from the remainder, or may receive different amounts of shade. These differences are unimportant so long as

Table 5.5   BASIC program for calculating additive effects of factorial designs.

```
5 DIM N(6),Y(72),E(24),C(6)
15 READ D
20 LET M=1
25 FOR I=1 TO D
30 READ N(I)
35 LET M=M*N(I)
40 NEXT I
45 FOR I=1 TO M
50 READ Y(I)
55 NEXT I
60 LET K=0
65 LET P=1
70 FOR I=1 TO D
75 LET C(I)=1
80 LET L=N(I)
85 LET P=P*L
90 LET K=K+L
95 NEXT I
100 FOR I=1 TO K
101 LET E(I+1)=0
105 NEXT I
110 LET T=0
115 FOR I=1 TO P
120 LET W=Y(I)
125 LET T=T+W
130 LET L=1
135 FOR J=1 TO D
140 LET K=L+C(J)
145 LET E(K)=E(K)+W
150 LET L=L+N(J)
155 NEXT J
160 LET J=D
165 IF C(J)=N(J) THEN 180
170 LET C(J)=C(J)+1
175 GO TO 195
180 LET C(J)=1
185 LET J=J-1
190 IF J<>0 THEN 165
195 NEXT I
200 LET R=1/M
205 LET E(1)=T*R
206 PRINT E(1)
210 LET K=1
215 FOR I=1 TO D
220 LET L=N(I)
225 LET W=L*R
230 FOR J=1 TO L
235 LET K=K+1
240 LET E(K)=(E(K)*W)-E(1)
245 PRINT E(K),
250 NEXT J
255 PRINT
260 NEXT I
265 STOP
400 END
```

all of the plots within a block are as similar as possible apart from the treatments applied to them. For all practical purposes, therefore, we can reduce our model to:

$$Y_j = \mu + \tau_j + \varepsilon_j$$

Table 5.6   BASIC program for tabulating additive effects for factorial designs.

```
5 REM PROGRAM TO CONSTRUCT ADDITIVE TABLE
10 DIM N(6),Y(72),E(24),C(6),V(6)
15 READ D
20 LET M=1
25 FOR I=1 TO D
30 READ N(I)
35 LET M=M+N(I)
40 NEXT I
45 FOR I=1 TO M
50 READ E(I)
55 NEXT I
265 LET P=1
270 LET K=1
275 LET A=E(1)
280 FOR I=1 TO D
285 LET L=N(I)
290 LET C(I)=L
295 LET P=P*L
300 LET K=K+L
305 LET V(I)=K
310 LET B=E(K)
315 LET A=A+B
320 LET J=K
325 IF L=1 THEN 350
330 LET E(J)=E(J-1)-E(J)
335 LET J=J-1
340 LET L=L-1
345 GO TO 325
350 LET E(J)=B-E(J)
355 NEXT I
360 LET T=P
365 LET Y(P)=A
370 LET I=D
375 LET A=0
380 LET J=V(I)
385 LET A=A+E(J)
390 LET C(I)=C(I)-1
395 LET V(I)=V(I)-1
400 IF C(I)<>0 THEN 425
405 LET C(I)=N(I)
410 LET V(I)=V(I)+C(I)
415 LET I=I-1
420 IF I=0 THEN 440
422 GO TO 380
425 LET T=T-1
430 LET Y(T)=Y(T+1)+A
435 GO TO 370
440 LET J=0
442 FOR I=1 TO P
445 PRINT Y(I),
446 LET J=J+1
447 IF J<>N(D) THEN 450
448 PRINT
449 LET J=0
450 NEXT I
470 STOP
600 END
```

where ε_j are random errors normally and independently distributed with mean zero and variance σ^2.

The technique of the analysis of variance is to partition the total sum of squares of deviations of the observed values from those

Table 5.7   Data and results for BASIC program of Tables 5.5 and 5.6.

```
NEW OR OLD--OLD
OLD PROGRAM NAME--ADDEFF

READY


   300 DATA 2,8,3
   301 DATA 0.84,1.09,1.16
   302 DATA 0.84,1.03,1.66
   303 DATA 0.97,0.88,1.50
   304 DATA 1.06,0.94,1.50
   305 DATA 1.00,1.13,1.59
   306 DATA 0.81,1.00,1.31
   307 DATA 0.75,1.06,1.44
   308 DATA 0.91,0.56,1.50

READY

RUN

 1.105417
 -.07541665      .07125001      .01125        .06125       .1345833
 -.06541666     -.02208333     -.1154167
 -.2079166      -.1441667       .3520833


READY

OLD

OLD PROGRAM NAME--ADDTAB

READY

500 DATA 2,8,3
501 DATA 1.105
502 DATA -0.075,0.071,0.011,0.061,0.135,-0.065,-0.022,-0.115
503 DATA -0.208,-0.144,0.352


RUN

 .8220001       .8860001      1.382
 .9680001      1.032         1.528
 .9080001       .9720001     1.468
 .9580001      1.022         1.518
1.032          1.096         1.592
 .832           .896         1.392
 .875           .939         1.435
 .782           .846         1.342


READY
```

expected by the underlying model. The partitioning reflects the number of parameters estimated, and provides a comparison of the additional variability introduced. The analysis appropriate to the data of Table 5.4 is summarized in Table 5.8. The total sum of squares of deviations is that derived by assuming that the expected value for each plot is the mean of the experiment. The sum of squares of

Table 5.8   Analysis of variance of seedling data of Table 5.4.

Source of variation	Degrees of freedom	Sum of squares of deviations	Mean square	F
Blocks	7	0·1525	0·0218	0·93
Treatments	2	1·5038	0·7519	32·00
Error	14	0·3925	0·0235	
Total	23	1·9858		

Mean squares are estimates of the following parameters:

Blocks	$\sigma^2 + t\sigma_\beta^2$
Treatments	$\sigma^2 + r\sigma_\tau^2$
Error	σ^2

deviations for blocks is that derived by calculating the expected value for each plot from the experimental mean and appropriate block effects, while the sum of squares of deviations for treatments is that derived by calculating the expected value of each plot from the experimental mean and the appropriate treatment effects. The error sum of squares of deviations, as might be expected, is derived by calculating the expected value of each plot from the experimental mean and the appropriate block and treatment effects. In practice, there are short-cut methods of calculating these sums of squares of deviations, and the calculations for this example are given in full by Jeffers.[38]

The term 'degrees of freedom' requires some explanation. In each of the three models summarized in Table 5.8, different numbers of parameters have been estimated. Although there are eight blocks, we have imposed the constraint that the block effects must add to zero, so that, in fact, only seven of those effects can be allocated independently—the eighth effect is then fixed, i.e. there are $7 = 8 - 1$ degrees of freedom. Similarly, because of the constraint of the model that the treatment effects must add to zero, there are only $2 = 3 - 1$ degrees of freedom in allocating treatment effects. The degrees of freedom for a model with both effects are obtained by multiplying the degrees of freedom for blocks and treatments, so that the degrees of freedom for the combined model are:

$$2 \times 7 = 14$$

Note that the sum of the degrees of freedom for the three models is equal to that for the total sum of squares of deviations, i.e. $23 = 24 - 1$, because we can give any values to 23 of the plots, but the value of the last plot is then constrained by the need to make the average expected value equal to the experimental mean.

The mean squares or variances are derived by dividing the sums of squares by deviations by the appropriate degrees of freedom. The mean squares of the three models represent the different composite variances also summarized in Table 5.8, and division of the mean squares for blocks and treatments by that for error provides a test of the significance of the block and treatment effects respectively, corresponding to the standard F-test of statistical literature. Comparison of the calculated ratios with tabulated values of F at the appropriate degrees of freedom gives the necessary indication of significance, and, in this case, so high a value for the treatment effects would not occur even once in a hundred times if the treatment effects were zero. The block effects, on the other hand, could easily have arisen from a population in which the block effects were in fact zero. For all practical purposes, therefore, our model reduces to:

$$Y_j = \mu + \tau_j + \varepsilon$$

the parameters of which are estimated by

$$Y_j = m + T_j + e$$

where $m = 1 \cdot 105$
$\quad\quad\quad T_1 = -0 \cdot 208$
$\quad\quad\quad T_2 = -0 \cdot 144$
$\quad\quad\quad T_3 = 0 \cdot 353$

and e is a random error normally and independently distributed with mean zero and variance $0 \cdot 0235/8 = 0 \cdot 00294$.

From this model, we may further estimate that the effect of formalin sterilization is:

$$d_{0,1} = (1 \cdot 105 - 0 \cdot 144) - (1 \cdot 105 - 0 \cdot 208) = 0 \cdot 064$$

with a standard error given by:

$$S_{\bar{x}} = \sqrt{2 \times \frac{0 \cdot 0235}{8}} = \pm 0 \cdot 077$$

Similarly, the effect of steam sterilization is estimated as:

$$d_{0,2} = (1 \cdot 105 + 0 \cdot 353) - (1 \cdot 105 - 0 \cdot 208) = 0 \cdot 561$$

again with a standard error of $\pm 0 \cdot 077$. The model, therefore, suggests an increase of $0 \cdot 561 \pm 0 \cdot 077$ mm in the mean diameter of the seedlings and that this increase in diameter is statistically significant. The increase of $0 \cdot 064 \pm 0 \cdot 077$ mm in the mean diameter of the seedlings produced by formalin sterilization is not statistically significant. The difference between the effects of steam and formalin sterilization is therefore of interest, and possibly the topic of further research.

The example of 5.2 is a very simple application of the linear additive model which underlies the analysis of variance, and anyone contemplating the use of systems analysis and mathematical models in ecology would be well-advised to study one of the general statistical texts dealing with the analysis of variance and covariance. The techniques developed in this form of analysis are extraordinarily powerful despite the apparent limitations of the basic assumptions which are:

1. The treatment and block effects are assumed to be additive.
2. The residual effects are assumed to be independent from observation to observation and to be distributed with zero mean and the same variance.
3. If tests of significance and estimated confidence limits are required, the residuals are assumed to be normally distributed.

Even where these assumptions can be regarded as only approximately true, or where the data have to be transformed to make the assumptions approximately true, the analysis of variance provides a method of constructing models of ecological populations, and of estimating the parameters of the models from sample observations. The models can be complex and contain linear and higher order interactions of many factors. For example, by carefully designed experiments and subsequent analysis, it is possible to test the effects of several different elements in fertilizer treatments, and simultaneously test whether the effects of each element are the same in the presence and absence of any or all of the other elements. Similarly, again by carefully designed experiments and the use of factorial models, the extent to which the management of chalk grassland by mowing and the use of fertilizer requires to be modified to take account of annual changes in climate can be determined. The development of additive models is, however, outside the scope of this introductory text.

MULTIPLE REGRESSION ANALYSIS

The linear models described above are a special case of the more general regression models characterized by the expression:

$$y = \beta_0 + \beta_1 x_1 + \beta_2 x_2 + \cdots + \beta_p x_p + \varepsilon$$

where β_0 is a constant
         β_i is the coefficient of the ith variable
         x_i is the ith variable

and   ε   are the random errors which are normally and independently distributed with mean zero and variance σ^2.

In this expression, y is assumed to be a random variable distributed
about a mean that is dependent on the values of the p variables
$x_1 \ldots x_p$. It is assumed that these variables affect only the mean of y,
and, in particular, that the variance is constant. Where tests of significance are required, it is further assumed that y is normally distributed
about this mean. Finally, it is assumed that the mean can be regarded
as a linear function of the x variables, although there can also be
functional relationships between the x's, so that polynomial and other
non-linear functions are included in these more general models.

As in the special case of the additive models of experimental
situations above, the parameters of the model are usually estimated by
minimizing the residual sum of squares $\sum (Y-y)^2$ for a sample from a
defined population. Sprent[75] gives a valuable discussion of the use of
models in regression, and advice and programs for the fitting of
regression models to actual data are given by Davies.[16] An example
of the application of a regression model will, however, give some idea
of the utility of such models in ecology.

Example 5.3 – Rates of decomposition of leaf litter.

In an experimental programme intended to determine the rates of
decomposition of tree and shrub leaf litter in glass tubes placed in field
conditions, but protected from animals, accurately weighed litter
samples of about 0·25 g air dry weight were allowed to decompose on
the surface of a mull soil in glass tubes 28 mm in diameter and 15 cm
long. Litter and soil were previously treated to kill animals and their
resting stages. The tubes were exposed in the field in boxes specially
designed to keep out animals, but to allow rain to pass through and air
to circulate. At intervals, the tubes were sampled randomly, two tubes
being taken at each sampling. Litter was weighed into specially designed
respiration flasks and respiration was measured as oxygen uptake over
a period of about eight hours after overnight equilibration in
respirometers.

The oxygen uptake was measured at three or four different temperatures, including 10°C, on successive days. The range of temperatures chosen covered the range found in the field at that time of
year. Preliminary investigations had shown that removing the litter
from the tubes did not produce a detectable difference in respiration,
and that results obtained at different temperatures on successive days
were the same when the temperature was initially low and was
increased, as when the temperature was initially high and was decreased.

After measurement of respiration, the litter was oven-dried at 105°C and weighed.

A full description of the analysis of the data from this experiment is given by Jeffers *et al.*[39] The values of oxygen uptake were plotted against temperature, and, from the resulting graph, it was clear that the relationship between oxygen uptake and temperature was not linear and that the variance of the oxygen uptake was greater at higher than at lower temperatures. The latter observation suggested the need for transforming the oxygen uptake values to their logarithms with the addition of a constant 1 to correct for zero uptakes, and the replotted data confirmed that the transformation adequately corrected for non-linearity and gave similar variances along the whole range of temperatures.

The equation:

$$\log_{10}(Y+1) = 0.561 - 8.701D \cdot 10^{-4} + 3.935D^2 \cdot 10^{-7}$$
$$+ 7.187M \cdot 10^{-4} + 0.0398T$$

where   Y is oxygen uptake measured in μl 0.25 g^{-1} h^{-1}
        D is the number of days for which the samples were exposed
        M is the per cent moisture content of the samples
and   T is temperature measured in °C

gave unbiased estimates of oxygen uptake over the full range of days, moisture contents and temperatures covered in this experiment, with an average deviation in oxygen uptake of 0.319 ± 0.321.

The use of the regression model gives considerable insight into the biological interpretation of the response of oxygen uptake to environmental factors. Two principal mathematical models are currently used by biologists to describe the relationship between the rate of a biological process and temperature. The first, and oldest, of these models is the Arrhenius equation, usually written in the exponential form involving the Boltzmann factor:

$$K = A \, e^{-E/RT}$$

where K is the specific reaction rate, A is a constant, E is a constant characteristic of the reaction and one which determines the influence of temperature on the reaction rate, R is the gas constant and T is the absolute temperature. Note that a logarithmic transformation of the equation gives

$$\log_e K = \log_e A + (-E/R)(1/T)$$

and the logarithm of the reaction rate is a linear function of the reciprocal of temperature.

The second model is associated with the temperature coefficient Q_{10}, which may be written as

$$Q_{10} = (K_2/K_1)^{10/(T_2 - T_1)}$$

When the temperature difference is $10°C$, the Q_{10} value is the ratio of the specific reaction rates at the two temperatures. This implies a relationship of the type

$$K = C\, e^{PT}$$

where C is a constant and $P = \log_e(Q_{10})/10$. Again, a logarithmic transformation of the relationship gives

$$\log_e K = \log_e C + P \cdot T$$

so that the logarithm of the reaction rate is a linear function of temperature.

The two models differ in that, for a given temperature difference, Q_{10} is independent of absolute temperature, whereas, for the Arrhenius equation,

$$K_2/K_1 = \exp\{-E/R(T_2^{-1} - T_1^{-1})\}$$

and, for a given temperature difference, this will vary with the absolute temperature.

The analysis of the oak litter supports the Q_{10} model, in that $P = \log_e(Q_{10})/10$ was independent of absolute temperature and equal to 0·09158, giving an estimated Q_{10} of 2·5.

MARKOV MODELS

The final type of stochastic model which we will consider in this introduction is that of the Markov models, which have a strong affinity to the matrix models discussed in Chapter 4. In these models, the basic format is of a matrix of entries expressing the probabilities of the transition from one space to another at specified intervals. The model is, therefore, exactly similar to those of the matrix models, except that all of the probabilities in the columns add to 1.

A first-order Markov model is one in which the future development of a system is determined by the present state of the system and is independent of the way in which that state has developed. The sequence of results generated by such a model is often termed a Markov chain. Application of the model to practical problems has three major limitations:

1. The system must be classified into a finite number of states.
2. The transitions must take place at discrete instants, although these

instants can be so close as to be regarded as continuous in time for the system being modelled.

3. The probabilities must not change with time.

Some modification of these constraints is possible, but at the cost of increasing the mathematical complexity of the model. Time-dependent probabilities can be used, as can variable intervals between transitions, and, in higher order Markov models, transition probabilities are dependent not only on the present state, but also on one or more preceding ones.

The potential value of Markovian models is particularly great, but has not so far been widely applied in ecology. However, preliminary studies suggest that, where ecological systems under study exhibit Markovian properties, and specifically those of a stationary, first-order Markov chain, several interesting and important analyses of the model can be made. For example,

1. Algebraic analysis of transition matrix will determine the existence of transient sets of states, closed sets of states or an absorbing state. Further analysis enables the basic transition matrix to be partitioned and several components investigated separately, thus simplifying the ecological system being studied.
2. Analysis of transition matrix can also lead to the calculation of the mean times to move from one state to another and the mean length of stay in a particular state once it is entered.
3. Where closed or absorbing states exist, the probability of absorption and the mean time to absorption can be calculated.

The ecological meaning of these terms is illustrated diagrammatically in Fig. 5.1, where a typical successional sequence is illustrated, each successional stage being described by the characteristics of the dominant types of plants present during that stage. The diagram shows that a transient set of states is one in which each state may eventually be reached from every other state in the set, but which is left when the state enters a closed set of states or an absorbing state.

A closed set differs from a transient set in that, once the system has entered any one of the states of the closed set, the set cannot be left. An absorbing state is one which, once entered, is not left: i.e. there is complete self-replacement. Mean passage time therefore represents the mean time required to pass through a specified successional state, and mean time to absorption is the mean time to reach a stable composition.

Extension of Markov models to second- and third-order levels, and to encompass limited degrees of non-stationarity and other forms of

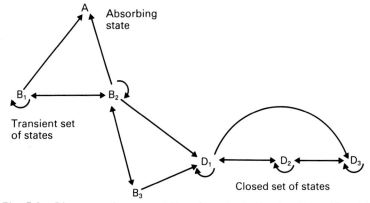

Fig. 5.1   Diagrammatic representation of transient, closed and absorbing states.

non-linearity is possible but considerably complicates subsequent analysis. Formal methods of testing the existence of second- and third-order dependence are also available. It is, however, relatively unlikely that such models will find wide application or acceptance in ecology in the near future.

To construct Markov-type models, the following main items of information are needed:

1. Some classification that, to a reasonable degree, separates successional states into definable categories (the multivariate models of
the next chapter are frequently useful in establishing such states).
2. Data to determine the transfer probabilities or rates at which states
change from one category of this classification to another through
time.
3. Data describing the initial conditions at some particular time,
usually following a well-documented perturbation.

The choice between Markov and other models often depends upon the objectives of the study, but, when a straightforward Markovian approach can be used, the possibilities of further algebraic analysis, leading to a better appreciation of the stochastic character of many ecological processes, as well as to the calculation of mean passage in time, time to absorption and the degree of stability and convergence within the defined state, provide additional information of direct ecological and management value.

The advantages of Markov-type models may therefore be briefly summarized as follows:

1. Such models are relatively easy to derive (or infer) from successional data.

2. The Markov model does not require deep insight into the mechanisms of dynamic change, but can help to pinpoint areas where such insight would be valuable, and hence acts both as a guide and a stimulus to further research.

3. The basic transition matrix summarizes the essential parameters of dynamic change in a system in a way which few other types of model achieve.

4. The results of the analysis of Markov models are readily adaptable to graphical presentation, and, in this form, are frequently more readily presented to, and understood by, resource managers.

5. The computational requirements of Markov models are modest, and can easily be handled on small computers, or, for small numbers of states, on hand calculators.

Markov models do, however, have some disadvantages, including the following:

1. The lack of dependence on functional mechanisms reduces their appeal to the functionally-orientated ecologist.

2. Departure from assumptions of stationary, first-order Markov chains in the case of straightforward Markov models, while conceptually possible, makes for disproportionate degrees of difficulty in analysis and computation.

3. In some cases, the data available will be insufficient to estimate reliable probabilities or transfer rates, especially for rare transitions.

4. As for other models, validation depends upon prediction of system behaviour which may be difficult for processes covering relatively long periods of time.

These difficulties warrant a high degree of caution in the uncritical application of Markov chains to ecological problems. Data collection, for the calculation of transition probabilities and the construction of transition matrices, is a major problem, ideally requiring detailed documentation of changes over periods of time, and response to various types of perturbations. However, where data of this type exist, either from historical or experimental records, such models are useful, and are likely to be more widely used in future.

Example 5.4 – Successional changes in a raised mire.

Raised mire frequently shows interesting successional changes as a result of increased drainage, and Table 5.9 gives estimated probabilities for the transitions between four possible states over a period of twenty years. State 1 represents the wettest facies dominated by *Sphagnum* with *Calluna vulgaris, Erica tetralix* and *Eriophorum vaginatum* as

Table 5.9   Transitional probabilities for successional changes in a raised mire (time step = 20 years).

Starting state	Probability of transition to end-state:			
	1. Bog	2. *Calluna*	3. Woodland	4. Grazed
1. Bog	0·65	0·29	0·06	0·00
2. *Calluna*	0·30	0·33	0·30	0·07
3. Woodland	0·00	0·28	0·69	0·03
4. Grazed	0·00	0·40	0·20	0·40

the major vascular plant components. State 2 represents a drier facies, with a *Calluna-Cladonia* association and seedlings of *Betula* and *Pinus sylvestris*. State 3 represents more or less established woodland of *Betula* and *Pinus sylvestris*, the more mature woodland having a typical *Vaccinium myrtillus* community with hypnaceous mosses. State 4 represents disturbance due to sporadic grazing by large herbivores of the drier facies, leading to the establishment of a *Molinia-Pteridium* dominated association.

Thus, areas which start as typical bog vegetation have a probability of 0·65 of remaining as bog vegetation at the end of twenty years, and probabilities of 0·29 and 0·06 respectively of becoming *Calluna*-dominated and woodland. Areas which start as *Calluna*-dominated have roughly equal probabilities of remaining in the same state, returning to bog vegetation because of fluctuations in the water-table, or of becoming woodland: they have a small (0·07) probability of being subjected to sporadic grazing. Woodland areas have a 0·69 probability of remaining woodland, a probability of 0·28 of returning to *Calluna* because of the death of trees, and, again, a small (0·03) probability of being subjected to sporadic grazing. The grazed areas have an equal probability of being subjected to continuous grazing and returning to a *Calluna*-dominated vegetation, and a smaller (0·20) probability of becoming woodland because of the growth of ungrazed seedlings.

None of the states, therefore, are absorbing or closed, but represent a transition from the bog vegetation to woodland, with an imposed disturbance due to grazing. However, although there can be a return from the *Calluna*-dominated vegetation to bog vegetation because of fluctuations in the water table, there is no immediate return to bog vegetation from woodland. Where there are no absorbing states, the Markov process is known as an ergodic chain and we can explore the full implications of the matrix of transition probabilities by exploiting the basic properties of the Markovian model.

First, as in the matrix models discussed in Chapter 4, the probabilities of Table 5.9 show the probability of the transitions from any one state to any other state after one time step (20 years). The transition probabilities after two time steps can be derived directly by multiplying the one-step transition matrix by itself, so that, in the simplest, two-state case the corresponding probabilities would be given by the matrix:

$$\begin{bmatrix} p_{11}^{(2)} & p_{12}^{(2)} \\ p_{21}^{(2)} & p_{22}^{(2)} \end{bmatrix} = \begin{bmatrix} p_{11} & p_{12} \\ p_{21} & p_{22} \end{bmatrix} \times \begin{bmatrix} p_{11} & p_{12} \\ p_{21} & p_{22} \end{bmatrix}$$

In condensed form, we may write:

$$\mathbf{P}^{(2)} = \mathbf{P}.\mathbf{P}$$

Similarly, the three-step transition may be written as:

$$\begin{bmatrix} p_{11}^{(3)} & p_{12}^{(3)} \\ p_{21}^{(3)} & p_{22}^{(3)} \end{bmatrix} = \begin{bmatrix} p_{11}^{(2)} & p_{12}^{(2)} \\ p_{21}^{(2)} & p_{22}^{(2)} \end{bmatrix} \times \begin{bmatrix} p_{11} & p_{12} \\ p_{21} & p_{22} \end{bmatrix}$$

or

$$\mathbf{P}^{(3)} = \mathbf{P}^{(2)}.\mathbf{P}$$

In general, for the nth step, we may write:

$$\mathbf{P}^{(n)} = \mathbf{P}^{(n-1)}.\mathbf{P}.$$

For the matrix of Table 5.9, the transition probabilities after two time-steps are:

$$\begin{bmatrix} 0\cdot5095 & 0\cdot3010 & 0\cdot1674 & 0\cdot0221 \\ 0\cdot2940 & 0\cdot3079 & 0\cdot3380 & 0\cdot0601 \\ 0\cdot0840 & 0\cdot2976 & 0\cdot5661 & 0\cdot0523 \\ 0\cdot1200 & 0\cdot3480 & 0\cdot3380 & 0\cdot1940 \end{bmatrix}$$

and after four time-steps are:

$$\begin{bmatrix} 0\cdot3648 & 0\cdot3035 & 0\cdot2893 & 0\cdot0424 \\ 0\cdot2759 & 0\cdot3048 & 0\cdot3649 & 0\cdot0543 \\ 0\cdot1841 & 0\cdot3056 & 0\cdot4528 & 0\cdot0595 \\ 0\cdot2151 & 0\cdot3114 & 0\cdot3946 & 0\cdot0789 \end{bmatrix}$$

If a matrix of transition probabilities is successively powered until a state is reached at which each row of the matrix is the same as every other row, forming a fixed probability vector, the matrix is termed a regular transition matrix. The matrix gives the limit at which the probabilities of passing from one state to another are independent of the starting state, and the fixed probability vector **t** expresses the equilibrium proportions of the various states. For our example, the vector of probabilities is:

$$[0.2177 \quad 0.2539 \quad 0.3822 \quad 0.1462]$$

If, therefore, the transition probabilities have been correctly estimated and remain stationary, the raised mire will eventually reach a state of equilibrium in which approximately 22 per cent of the mire is bog, and approximately 25 per cent, 38 per cent, and 15 per cent are *Calluna*, woodland and grazed communities respectively.

Where, as in this example, there are no absorbing states, we may also be interested in the average lengths of time for an area of bog to become *Calluna*-dominated, woodland, or grazed—i.e. the mean first passage times. Alternatively, if we choose an area at random, what is the average lengths of time we would need to wait for this area to become bog, *Calluna*, woodland, or grazed—i.e. the mean first passage times in equilibrium?

The computations are relatively complex, but the matrix of mean first passage times is:

$$\begin{bmatrix} 0 & 3.561 & 7.197 & 31.688 \\ 9.566 & 0 & 5.237 & 28.755 \\ 13.672 & 4.107 & 0 & 29.178 \\ 18.673 & 9.107 & 5.000 & 0 \end{bmatrix}$$

As each time-step represents twenty years, the average length of time for a *Calluna*-dominated area to become bog is $9.566 \times 20 = 191$ years. Similarly, the average length of time needed for a woodland to become *Calluna* is $4.107 \times 20 = 82$ years, and the other times can be calculated as required.

Finally, the mean first passage times in equilibrium are given by the vector:

$$[10.385 \quad 3.676 \quad 3.627 \quad 25.351]$$

Again, as each time-step represents twenty years, the mean first passage time for a randomly chosen area to become bog is $10.385 \times 20 = 208$ years, while the corresponding mean first passage times for *Calluna*-dominated, woodland and grazed communities are 74 years, 73 years, and 507 years respectively.

As with many of the other forms of mathematical models, the basic properties of the model provide further information about the behaviour of the systems of which they are the model. We are exploiting known relationships for mathematical models and thereby avoiding tedious experimentation which would otherwise be necessary to determine the properties of an empirical dynamic model.

In the historical development of mathematical models, the deterministic models were the first to be developed. Most of these models, however, were developed in relation to applications of physical and chemical laws, studied under conditions in which variability of response was relatively small or, at least, relatively easy to control. These physical analogues perhaps came to their fullest flowering in the development of the Newtonian calculus, although there has continued to be a steady and impressive development of the underlying mathematics of systems which are deterministic. The dynamic models described in Chapter 3 represent the first attempt to apply this deterministic thinking to biological and ecological situations. As stressed in Chapter 1, ecological relationships necessarily involve the inherent variability of organisms and of habitats, as well as variability in the interaction between those organisms and those habitats. If, therefore, we are to model that inherent variability, some appropriate stochastic relationships also have to be invoked.

In part, the recognition that biological entities have an inherent variability was one of the mainsprings of the development of modern statistics. Born of the desire to predict the outcome of games of chance and gambling, the studies of probabilities developed in the eighteenth and nineteenth centuries were further developed in ingenious and enlightening ways by the new schools of statisticians seeking analogues for biological processes and for the modelling of agricultural and forestry experiments. This development of statistics, too, has been continued to what can only be regarded as an astonishing degree, with the mathematics becoming both more elegant and more revealing of the basic properties of the models. In this chapter, we can only give a glimpse of some of the advantages of using these well-developed families of mathematical models to represent processes of interest in both research and management. As in the last chapter, we accept greater formalization of the models for the sake of the easier understanding of the properties of the model, although we are never absolved from the necessity of testing that the assumptions made by the mathematics are valid for the particular problem in which we are interested. We will return to this topic in the last chapter of this book.

There are two ways in which deterministic models fail to mirror ecological reality. First, they assume infinite population sizes, and second, they ignore random fluctuations in the environment with time. Hence, the use of deterministic rather than stochastic models can only be justified by mathematical convenience. For example, if a deterministic model shows a stable equilibrium, the corresponding stochastic model will almost certainly predict long-term survival, whereas, if a deterministic model shows no equilibrium or an unstable one, the stochastic model will usually predict extinction with a high probability. However, while we may expect the deterministic and stochastic models to have similar properties, it is frequently the contrast between the predictions made from seemingly valid alternative models which is of greatest interest. It is for this reason that the method of systems analysis places particular emphasis on the simultaneous investigation of several alternative models as a solution to a practical problem. Where a deterministic model is used, it is probably always wise to balance the results from that model against those from one or more stochastic models. However, even when the deterministic model is the only one to be adopted, it will usually be necessary to estimate the parameters of that model by an appeal to the mathematics of statistics and stochastic processes. These methods are, therefore, of fundamental importance to the application of systems analysis to ecology.

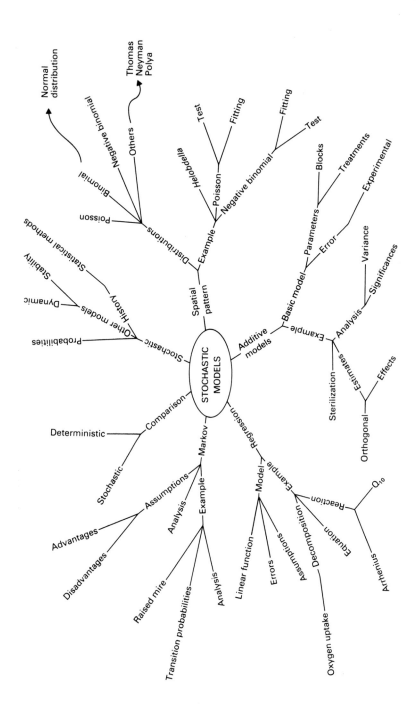

6

Multivariate Models

So far, in this text, we have tended to use the term 'variable' for any quantity which takes different values for different individuals, or different values for the same individual at different times. The statistical definition of the term 'variable' is 'a quantity which may take any one of a specified set of values', and these values may be continuous, as in measurements of height, or discontinuous, as in counts of individuals. Indeed, in some instances, it is convenient to use the word to denote non-measurable characteristics. For example, 'sex' may be regarded as a variable in this sense, as any individual may take one of two values, 'male' or 'female'.

In Chapter 2, we made a distinction between dependent, and independent or regressor variables. Dependent variables are those which may be expected to be altered by changes in other variables, while regressor variables describe those variables which provide the changes necessary to induce changes in the dependent variable. We have already encountered the distinction between these two classes of variables in the regression models of the last chapter.

A statistician makes a further distinction between variables and variates. A variate is a quantity which may take any one of the values of a specified set with a specified relative frequency or probability. Such variates are sometimes also known as random variables and they are to be regarded as defined, not merely by a set of permissible values like any ordinary mathematical variable, but by an associated frequency or probability function expressing how often those values appear in the situation under discussion. Almost all of the models which we have so far considered have been defined in terms of only one variate, but there are many situations in ecology and other applications of systems analysis, where models have to capture the behaviour of more than one variate. These models are known collectively as 'multivariate' and are related to techniques known collectively as 'multivariate analysis'—an expression which is used rather loosely to denote the analysis of data which are multivariate in the sense that each member bears the values of p variates.

Much of the mathematics involved in the construction of multivariate models is not new. For example, the fundamental probability distributions related to the multivariate normal distributions were derived in the 1930's, and the methods developed then are the basis of most of the multivariate methods used today. However, the calculations involved in multivariate analysis and in the construction of multivariate models become extremely onerous when the number of variates is large, and these calculations were virtually impossible to carry out, even on electric calculating machines, for more than four or five variables. Until the electronic computer became generally available, therefore, only a very limited number of multivariate analyses had been attempted, and the same examples were therefore quoted in almost all of the text books.

The availability of electronic computers has now completely changed the situation, with the result that multivariate models have become an important addition to the range of models with which systems analysis may be concerned. Most of the computations have been programmed for the range of computers now available, and a rapidly increasing collection of examples has appeared in the scientific literature. Nevertheless, these multivariate models represent a much-neglected class of models in the application of systems analysis to ecology and other topics.

Figure 6.1 gives a simple classification of multivariate models. Broadly, these models may be divided into two main categories, i.e. those in which some variates are used to predict others and those in which all the variates are of the same kind, and no attempt is made to predict one set from the other. For the latter, which may be broadly described as descriptive models, there is a further subdivision into those models in which all the inputs are quantitative and which include principal component analysis and cluster analysis, and those models for which some at least of the inputs are qualitative rather than quantitative. For the latter, the reciprocal averaging model is the more appropriate. Predictive models, on the other hand, may first be subdivided according to the number of variates predicted and then by whether or not the predictors are all quantitative. Where several variates are predicted, the model of canonical analysis is the most

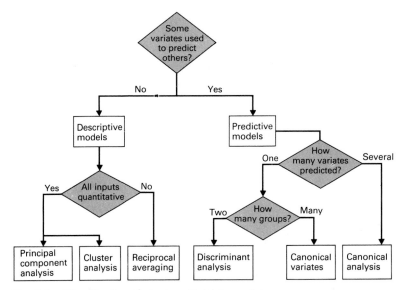

Fig. 6.1 Classification of some multivariate models.

appropriate. Where only one variate is predicted and there are two
a priori groups of individuals, the model of discriminant analysis is the
most appropriate of the available models, while where there are more
than two *a priori* groups of individuals canonical variate analysis will
provide the most useful approach. Simple examples of these various
models are described in the remainder of this chapter.

DESCRIPTIVE MODELS

Principal component analysis

Principal component analysis is probably the best known of the
multivariate models, and is certainly one of the simplest ways of
studying multivariate variation. It is a technique which can be applied
to all data which satisfy the following basic requirements:

1. For each of a number of individual sampling units, the same
   variables are measured and recorded. All the measurements must
   be made on each individual, and any individual for which the
   measurements are incomplete should be eliminated from the analysis, unless some suitable technique for replacing the missing values
   is found.

2. The variables selected for analysis are assumed to be continuous, or, if discrete, to increase by sufficiently small intervals of measurement as to be regarded as approximately continuous. It is possible that the analysis can be extended to deal with qualitative attributes which are scaled or scored, but only by weakening some of the basic assumptions, and alternative methods of analysis are therefore preferable.
3. No attempt should be made to add ratios or linear functions of the original variables to those to be included in the analysis or to replace any of the originally measured variables by ratios or linear functions.

The objectives of a principal component analysis may include one or more of the following:

1. Examination of the correlations between the separate variables.
2. Reduction of the basic dimensions of the variability expressed by the individual sampling units to the smallest number of meaningful dimensions.
3. Elimination of variables which contribute relatively little extra information to the study.
4. Examination of the most informative groupings of the individual sampling units, or the implications of some *a priori* structure imposed upon the sample units.
5. Determination of the objective weighting of the variables in the construction of indices of variation.
6. Identification of individual sampling units of doubtful or unknown origin.
7. The recognition of mis-identified sample units.

Not all of these objectives will be of equal importance in any particular study, and some may be entirely absent. Nevertheless, the method provides one possible solution to such problems.

Principal component analysis is described in detail by Kendall,[41] Quenouille[64] and by Seal.[70] Applications of the technique to ecological problems, and particularly those of taxonomy, are now relatively frequent, and most issues of ecological journals will contain at least one application of this technique.

In essence, principal component analysis involves the extraction of the eigenvalues and eigenvectors of the matrix of correlation coefficients of the original variables. The resulting eigenvalues and eigenvectors define the components of the total variability described by the original variables as linear functions of these variables with coefficients so chosen that the functions are mathematically independent, or orthogonal, to each other. There is no necessity that the components have any valid ecological interpretation, but practical experience with the technique suggests that a valid interpretation may usually be expected for those components which account for a significant part of the total variation. Furthermore, the calculation of the value of these components for each of the individuals included in the study provides a ready way of summarizing the essential variation of the individuals, examining their relationships with each other, and identifying unknown or misplaced individuals. The technique is best illustrated by a simple example.

Example 6.1 – Physical environment and invertebrates in Morecambe Bay.

In the course of a major investigation of the likely environmental impact of the construction of a barrage across the mouth of Morecambe Bay, a survey was undertaken during the months of August and September 1968. Ten 10 cm cores were taken from each of 274 sampling points at various parts of the Bay, the sampling material being bulked immediately, and one quarter of the bulked sample being retained for chemical and physical analysis. Eight variables were assessed for each of the samples, the variables being as follows:

1. Percentage of particles >250 μm
2. Percentage of particles $125–250$ μm
3. Percentage of particles $62\cdot5–125$ μm
4. Percentage of particles $<62\cdot5$ μm
5. Percentage loss on ignition at $550°C$
6. Percentage calcium
7. Percentage phosphorus
8. Percentage nitrogen.

The choice of these variables was guided by previous analysis of data from a pilot survey, so that the number of variables in the main survey has already been guided by the application of principal component analysis.

The basic data for the 274 sampling points are summarized in Table 6.1, and the coefficients of correlation between the original variables are given in Table 6.2. Application of the usual test of significance of the correlation between two variables, with degrees of freedom $n - 2 = 274 - 2 = 272$ (a test of doubtful validity for a correlation matrix of this kind), indicated that the percentages of the particles greater than 250 μm and between 125–250 μm were significantly positively correlated, and were negatively correlated with the per-

Table 6.1   Summary of environmental variables for Morecambe Bay.

Variable (%)	Minimum	Mean	Maximum	Standard deviation
1. Particles >250 μm	0·100	1·207	43·0	4·479
2. Particles 125–250 μm	0·050	20·31	94·0	23·27
3. Particles 62·5–125 μm	0·100	53·67	97·0	21·36
4. Particles <62·5 μm	0·500	24·74	88·0	20·77
5. Loss on ignition at 550°C	0·440	1·504	3·72	0·555
6. Calcium	1·500	2·401	9·00	0·704
7. Phosphorus	0·016	0·028	0·048	0·0056
8. Nitrogen	0·001	0·013	0·054	0·0093

centages of particles between 62·5–125 μm and below 62·5 μm. The percentages of particles between 62·5–125 μm and below 62·5 μm were also significantly negatively correlated. In contrast, the four chemical variables were all significantly and positively intercorrelated. Loss on ignition was positively correlated with the percentages of particles between 62·5–125 μm and below 62·5 μm, and negatively correlated with the percentage of particles between 125–250 μm. The percentage calcium was positively correlated with the percentages of particles greater than 250 μm, and less than 62·5 μm, and negatively correlated with the percentage of particles between 125 and 250 μm. Phosphorus content was negatively correlated with the percentages of particles above 125 μm and positively correlated with the percentages of particles below 125 μm. The nitrogen content was negatively correlated with the percentage of particles between 125–250 μm and positively correlated with the percentage of particles below 62·5 μm.

Apart from the fact that there are very considerable intercorrelations between these eight variables, the interpretation of the relationships is far from easy to determine by casual inspection of the correlation matrix, or, indeed, of the data themselves. Principal component analysis begins by computing the linear function of the eight variables which will account for as much of the variation contained by the 274 samples as possible, and defines this as Z_1. The analysis then goes on to identify a second linear function of the original variables which is independent of the first and accounts for as much as possible of the residual variability. This linear function is designed as Z_2. The analysis then continues to find Z_3, Z_4, etc., until all the variability has been accounted for. In practice, the computation is equivalent to extracting the eigenvalues and eigenvectors of the correlation matrix in Table 6.2 (Krzanowski;[44] Kendall;[41] Kendall and Stuart;[42] Seal[70]).

Table 6.2 Coefficients of the correlations between environmental variables.

x_1							
0·147†	x_2						
−0·283†	−0·565†	x_3					
−0·095	−0·572†	−0·330†	x_4				
−0·001	−0·462†	0·127*	0·388†	x_5			
0·713†	−0·253†	−0·051	0·175†	0·359†	x_6		
−0·148*	−0·405†	0·217†	0·264†	0·566†	0·167†	x_7	
0·072	−0·426†	0·005	0·453†	0·735†	0·421†	0·436	x_8

* significant at 0·05.   † significant at 0·01.

$$Z_1 = C_{11}X_1 + C_{21}X_2 + C_{31}X_3 + C_{41}X_4 + C_{51}X_5 + C_{61}X_6 \\ + C_{71}X_7 + C_{81}X_8$$

where Z_1 = the first principal component
  C_{ij} = the coefficient of the ith variable for the jth component
  X_i = the ith variable

$$Z_2 = C_{12}X_1 + C_{22}X_2 + C_{32}X_3 + C_{42}X_4 + C_{52}X_5 + C_{62}X_6 \\ + C_{72}X_7 + C_{82}X_8$$

$$Z_i = C_{1i}X_1 + C_{2i}X_2 + C_{3i}X_3 + C_{4i}X_4 + C_{5i}X_5 + C_{6i}X_6 \\ + C_{7i}X_7 + C_{8i}X_8$$

The first four components of the correlation matrix of Table 6.2 are summarized in Table 6.3. The eigenvalue of the first component is 3·12, and this eigenvalue, expressed as a percentage of the total number of variables, indicates the proportion of the total variability accounted for by that component. Similarly, the remaining eigenvalues summarize the proportion of variability accounted for by the appropriate component and these proportions can be added to give the cumulative proportions of the total variability accounted for by the linear functions which are, by definition, independent. From Table 6.3, therefore, we can see that the first linear function of the eight variables accounts for 39 per cent of the total variability, and the next three components for

Table 6.3 Eigenvalues of the first four components of the environmental variables.

Component	Eigenvalue	Proportion of variability	Cumulative proportion
z_1	3·12	39·0	39·0
z_2	1·87	23·4	62·4
z_3	1·26	15·7	78·1
z_4	0·83	10·3	88·4

Table 6.4 Eigenvectors of the first four components of the environmental variables.

| Variables | Coefficients for component: | | | |
	z_1	z_2	z_3	z_4
x_1	0·05	1·00	0·49	0·17
x_2	−0·90	0·40	−0·23	−1·00
x_3	0·25	−0·72	1·00	0·24
x_4	0·74	0·07	−0·87	0·84
x_5	1·00	0·01	−0·03	−0·64
x_6	0·61	0·79	0·53	0·24
x_7	0·80	−0·27	0·04	−0·86
x_8	0·97	0·17	−0·16	−0·42

23·4, 15·7 and 10·3 per cent respectively. The four components together account for 88·4 per cent of the variability measured by the eight original variables. The analysis also indicates that it is probably not worth computing any further components, components with eigenvalues of less than approximately o·8 being unlikely to have any practical value, although the reader should note that we are not making any appeal to the usual type of significance test. The reason for the absence of explicit significance testing is that the only known tests of significance for multivariate problems depend upon the assumption of a multivariate normal distribution which is seldom justified in ecological data.

The coefficients of the linear functions defining the components, and estimated by the eigenvectors, are summarized in Table 6.4. We may use these coefficients to interpret the ecological meaning of the components, using the sign and relative size of the coefficient as an indication of the weighting to be placed upon each variable in the four indices of variability. The first component is essentially a contrast of the loss on ignition and percentages of phosphorus and nitrogen with the percentage of particles between 125–250 μm, and represents a measure of the general 'fertility' of the sands and muds. The second component is an index of the percentage of the largest particles, i.e. those >250 μm, and the calcium content, and is a measure of the amount of broken shells of biological organisms. The third component is a contrast of the percentage of particles between 62·5–125 μm with the percentage of particles <62·5 μm, and is interpreted as a measure of the deposition of sea-borne silt. The fourth component is again a contrast, but this time between the percentage of particles between 125–250 μm and the phosphorus content with the percentage of particles below 62·5 μm, and is interpreted as a measure of river-borne deposits.

Table 6.5   Summary of species counts for Morecambe Bay.

| Species | Numbers per square metre | | | Standard deviation |
	Minimum	Mean	Maximum	
y_1 *Macoma balthica*	0	2325	56 325	5966
y_2 *Tellina tenuis*	0	49·2	9800	544
y_3 *Hydrobia ulvae*	0	374·2	8525	1014
y_4 *Corophium volutator*	0	540·5	8700	1180
y_5 *Nereis diversicolor*	0	63·5	750	116
y_6 *Arenicola marina*	0	16·7	222	26
y_7 *Nephthys hombergii*	0	4·94	100	17

The analysis suggests that a limited number of dimensions is sufficient to account for the major variability in the chemical and physical properties of the sands and muds of Morecambe Bay. In this case, four components were sufficient to account for 88 per cent of the total variability, and the resulting components are readily interpreted in terms of identifiable types of variation. Indeed, calculation of the values of the individual components for individual samples, and the plotting of these values on Morecambe Bay, helps to identify areas of high fertility, sea-borne deposits, and river-borne deposits, as well as those areas where a high calcium content indicated the presence of broken shells of estuarine organisms. The resulting maps greatly aid the interpretation of what would otherwise remain a somewhat incomprehensible source of variation.

In conjunction with the survey of particle size and chemical composition of the sands and muds, samples were also examined to determine the numbers of some 22 species or species groups of invertebrates. Table 6.5 summarizes the numbers per square metre of seven of the species, the other species being too sparsely distributed for analysis to be worthwhile. The total number of samples included in this table was 329, some additional samples to those for determining the environmental variables being sampled for the distribution of species, as these were thought to be more variable than the environmental variables. Inspection of Table 6.5 suggests that this assumption was certainly true, there being an extremely wide variation in the number of individual organisms of the seven species in the mud samples.

The coefficients of the correlations between the numbers of species in the individual samples are summarized in Table 6.6. Again, the usual test of significance for the coefficients of the correlation between two variables is scarcely relevant, not only because we are here testing several coefficients simultaneously, but also because the original counts are far from normally distributed. However, under the usual

Table 6.6   Coefficients of the correlations between species numbers.

y_1						
−0·028	y_2					
0·358†	0·032	y_3				
0·051	0·054	0·313†	y_4			
0·569†	0·009	0·302†	0·162†	y_5		
0·174†	−0·003	0·081	−0·095	0·084	y_6	
−0·170†	−0·000	−0·099	−0·118*	−0·092	−0·011	y_7

* significant at 0·05.   † significant at 0·01.

test, the numbers of *Macoma balthica* were positively correlated with those of *Hydrobia ulvae*, *Nereis diversicolor*, and *Arenicola marina* and negatively correlated with those of *Nephthys hombergii*.

The numbers of *Hydrobia ulvae*, *Corophium volutator*, and *Nereis diversicolor* were all intercorrelated, and the numbers of *Corophium volutator* negatively correlated with those of *Nephthys hombergii*. The numbers of *Tellina tenuis* were not markedly correlated with those of any other organisms.

As before, the principal components of the correlation matrix of Table 6.6 are defined by the eingenvalues and eigenvectors of that matrix. Table 6.7 summarizes the first five eigenvalues of the correlations between the species counts, and suggests that these first five components account for nearly 86 per cent of the total variability contained in the species counts. The remaining two components that can be extracted probably indicate only random variation. From the eigenvectors given in Table 6.8, the first component, accounting for 28·3 per cent of the variability in species numbers, is an index of the numbers of *Macoma balthica*, *Hydrobia ulvae* and *Nereis diversicolor*. The second component, accounting for a further 17·1 per cent of the variability, is a contrast of the numbers of *Corophium volutator* with those of *Arenicola marina*. The remaining components, accounting for 14·3, 13·6 and 12·2 per cent respectively, are direct measures of the numbers of *Tellina tenuis*, *Nephthys hombergii* and *Arenicola marina* respectively.

Table 6.7   Eigenvalues of the first five components of the species counts.

Component	Eigenvalue	Proportion of variability	Cumulative proportion
w_1	1·98	28·3	28·3
w_2	1·20	17·1	45·4
w_3	1·00	14·3	59·7
w_4	0·95	13·6	73·3
w_5	0·85	12·2	85·5

Table 6.8   Eigenvectors of the first five components of the species counts.

| Variable | Coefficients for component: | | | | |
	w_1	w_2	w_3	w_4	w_5
y_1	1·00	−0·48	−0·01	0·13	−0·43
y_2	0·04	0·40	1·00	−0·20	−0·41
y_3	0·89	0·32	0·04	0·11	0·59
y_4	0·51	1·00	−0·05	0·13	0·71
y_5	0·99	−0·22	−0·01	0·19	−0·59
y_6	0·29	−0·86	0·25	−0·54	1·00
y_7	−0·32	−0·44	0·34	1·00	0·47

Again, the analysis succeeds in summarizing a complex mass of information in a relatively simple way, the five components accounting for nearly 86 per cent of the total variability. As for the chemical and physical properties of the sands and muds, calculation of the values of the individual components for individual samples, and the plotting of these values on a map, gives a comprehensive picture of the distribution of the organisms within these five independent components. The resulting maps again aid the interpretation of what would otherwise remain a somewhat incomprehensible and uncharacterized source of variation. Even further interest is generated, however, by examining the correlations between the computed values of the invertebrate and physical and chemical components for the 272 samples for which both sets of components were available. These correlations are summarized in Table 6.9. Again, with reservations about the validity of the significance tests between such correlations, the table suggests some interesting relationships between the environmental and invertebrate components. The first invertebrate component, an index of the numbers of *Macoma balthica*, *Hydrobia ulvae* and *Nereis diversicolor*, is positively correlated with the first of the environmental components, an index of general fertility. The contrast between the numbers of *Corophium volutator* and *Arenicola marina* is negatively correlated

Table 6.9   Coefficients of correlations between environmental and invertebrate components.

| Invertebrate component | Correlation coefficient with environmental component: | | | |
	z_1	z_2	z_3	z_4
w_1	0·408†	−0·029	0·039	−0·031
w_2	−0·047	−0·164†	−0·097	0·153*
w_3	−0·078	−0·007	0·120*	0·000
w_4	−0·029	−0·187†	0·062	−0·101
w_5	−0·004	0·042	0·095	0·156†

* significant at 0·05.   † significant at 0·01.

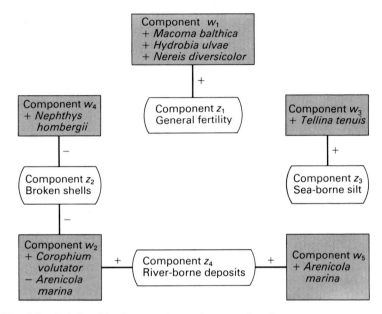

Fig. 6.2   Relationships between invertebrates and environment.

with the second environmental component and positively correlated with the fourth component, i.e. negatively with the presence of broken shells, and positively with the deposit of river-borne material.

The numbers of *Tellina tenuis* are positively correlated with the third environmental component, which is a measure of the deposition of sea-borne silt, while the numbers of *Nephthys hombergii* are negatively correlated with the presence of broken shells. The numbers of *Arenicola marina* are positively correlated with the deposits of sea-borne material. These correlations are further summarized in Fig. 6.2. The two analyses provide an interesting and descriptive analysis of the relationships between the physical and chemical properties of the sands and muds of Morecambe Bay and the invertebrate populations. We will, however, return to this example when we come to consider the alternative model of the canonical correlation.

Cluster analysis

An alternative multivariate descriptive model, when all inputs are quantitative, is that of cluster analysis. Cluster analysis encompasses many diverse techniques for discovering structure within complex bodies of data. In a typical example, and as in principal component analysis, the data base consists of a sample of units each described by a series of selected variables. The objective is to group either the data units or the variables into clusters so that the elements within a cluster have a high degree of 'natural association' among themselves, while the clusters are 'relatively distinct' from one another. The approach to the problem and the results achieved depend principally on how the investigator chooses to give operational meaning to the phrases 'natural association' and 'relatively distinct'.

In general, cluster analysis assumes that little or nothing is known about the structure which underlies a data set. All that is available is the collection of observations whose structure is unknown. The operational objective in this case is to discover a category structure which fits the observation, and the problem is frequently stated as one of finding 'natural groups'. The essence of cluster analysis might equally be viewed as that of assigning appropriate meaning to the terms 'natural groups' and 'natural associations'.

Cluster analysis is the attempt to group sample points in multidimensional space into separate sets which, it is hoped, will correspond to observed features of the sample. The groups of points may themselves be grouped into larger sets, so that all the points are eventually classified hierarchically. This hierarchical classification can be represented diagrammatically, and it is usual to incorporate a scale into such a diagram to indicate the degree of similarity of the various groups. One of the simplest forms of cluster analysis is a single linkage cluster analysis, a method suggested by Sneath[74] as a convenient way of summarizing taxonomic relationships in the form of dendrograms. The relationships between n samples are expressed in terms of taxonomic distances, measured on some acceptable scale, between every pair of samples. The method consists of a way of sorting the samples that determines clusters for a series of increased distance thresholds $(d_1, d_2, \ldots, d_n)$. The clusters at any level d_i are constructed from the following activities:

(a) The samples are grouped by joining all segments of length d_i or less. Each set is said to form a cluster at level d_i, and all segments joining two clusters defined at level d_i will have lengths greater than d_i.

(b) If sorting is done at a greater distance threshold d_{i+1}, all clusters at level d_i remain, but some of these clusters may combine into larger clusters. In general, two clusters will combine when at least one link exists between them of length d, where $d_i < d \leqslant d_{i+1}$. (This property of requiring only one link for the combination of groups explains the name 'single linkage cluster analysis'.)

The dendrogram shows how clusters at level d_1 combine at level

d_2, and so on, at successive levels, until all samples combine into a single cluster. In practice, the single linkage cluster analysis can be conveniently derived from what is known as the minimum spanning tree, i.e. a tree spanning all points by a set of straight line segments joining pairs of points, such that:

   (i) no closed loops occur
  (ii) each point is visited by at least one line
 (iii) the tree is connected
 (iv) the sum of the length of the segments is a minimum.

Figure 6.3 is a simple example of a tree with integer segment lengths, and a total length of 22.

    Cormack[14] has reviewed the various techniques for cluster analysis which have been developed during recent years, and has described the principles of the many empirical classification techniques, and the limitations and shortcomings in their development and practice. In general, some caution should be used in attempting cluster analysis of data, and methods should be based on well-defined mathematical formulations of the problem. The growing tendency to regard classification and cluster analysis as a satisfactory alternative to clear thinking needs to be condemned, and other ways of summarizing data may often be suggested as alternatives to cluster analysis and classification itself. Nevertheless, when cluster analysis is used as only one of the models of a systems analysis, the results may be revealing and helpful. Again, the technique is best described by a simple example.

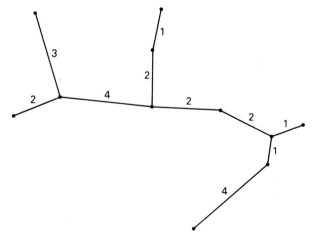

Fig. 6.3   Example of a minimum spanning tree with integer values.

Table 6.10 Values of seven variables for 25 Lake District soils.

Soil no.	Loss on ignition (% dry wt.)	Isotopically exchangeable phosphorus (μg g^{-1} dry wt.)	Phosphatase activity*	Extractable iron (mg 100 g^{-1} dry wt.)	Total phosphorus (% dry wt.)	Total nitrogen (% dry wt.)	pH
1	15·21	70·6	467·1	1400	0·12	0·63	4·53
2	33·27	67·5	1059·8	460	0·15	1·19	4·90
3	68·09	1700·3	3309·7	1200	0·36	2·30	4·82
4	32·89	168·1	1392·9	2100	0·17	1·29	4·84
5	19·87	102·7	71·3	920	0·14	0·73	7·93
6	16·46	32·5	367·0	1100	0·06	0·52	3·78
7	10·56	192·9	352·4	1000	0·10	0·33	4·59
8	15·63	118·4	300·2	1900	0·11	0·61	4·16
9	11·15	101·4	308·4	1300	0·11	0·47	5·13
10	16·25	232·5	306·2	1600	0·12	0·66	4·43
11	9·94	51·4	212·3	1800	0·10	0·37	4·70
12	70·63	150·3	627·7	590	0·15	1·81	3·65
13	9·0	9·8	129·7	95	0·01	0·21	3·63
14	19·71	297·7	467·9	2200	0·08	0·63	4·04
15	26·02	83·9	618·3	2800	0·08	0·88	3·93
16	11·84	168·9	375·8	750	0·07	0·45	5·89
17	10·71	127·3	330·3	910	0·13	0·43	4·56
18	8·3	107·4	241·4	880	0·08	0·31	4·74
19	12·67	188·7	516·4	1300	0·05	0·33	4·40
20	15·92	203·6	336·9	1500	0·08	0·52	4·13
21	12·92	170·6	319·6	1600	0·06	0·44	4·05
22	7·54	53·8	315·7	890	0·05	0·28	4·70
23	21·96	104·3	578·8	1900	0·12	0·81	4·11
24	88·78	107·6	1156·8	290	0·06	1·99	3·19
25	72·19	174·7	1061·3	690	0·14	2·32	3·93

* Expressed as μg phenol g^{-1} dry wt. soil in 3 hours at 13°C.

Example 6.2 – Lake District soils.

As an example of cluster analysis based on the minimum spanning tree and single linkage, we will examine the properties of 25 soils in the Lake District used in a study of the responses of sycamore and birch to soil nutrition. The soils were chosen to give as wide a range as possible of chemical properties, especially in terms of phosphate nutrition. Before using the soils in the experiments to test the response of sycamore and birch from different provenances, it was necessary to investigate the range of variability of the soils and their possible groupings.

Table 6.11   Summary of soil data.

Variable	Minimum	Mean	Maximum	Standard deviation
Loss on ignition (% dry wt.)	7·54	25·50	88·78	23·26
Isotopically exchangeable phosphorus (μ g^{-1} dry wt.)	9·80	191·48	1700·30	321·37
Phosphatase activity (μg phenol g^{-1} dry wt. soil in 3 hours at 13°C)	71·3	608·96	3309·70	653·68
Extractable iron (mg 100 g^{-1} dry wt.)	95	1247	2800	644·44
Total phosphorus (% dry wt.)	0·01	0·108	0·36	0·065
Total nitrogen (% dry wt.)	0·21	0·820	2·32	0·634
pH (in water)	3·19	4·510	7·93	0·909

Table 6.10 gives the values of seven variables determined for each of the 25 soils, including loss on ignition, isotopically exchangeable phosphorus, phosphatase activity, extractable iron, total phosphorus, total nitrogen, and pH. The data are further summarized in Table 6.11. The Euclidean distance between any two soils is calculated from the formula:

$$d_{ij} = [(x_{1i}-x_{1j})^2 + (x_{2i}-x_{2j})^2 + (x_{3i}-x_{3j})^2 + (x_{4i}-x_{4j})^2$$
$$+ (x_{5i}-x_{5j})^2 + (x_{6i}-x_{6j})^2 + (x_{7i}-x_{7j})^2]^{1/2}$$
$$= \left| \sum_{k=1}^{7} (x_{ki}-x_{kj})^2 \right|^{1/2}$$

where d_{ij} is the Euclidean distance between the ith and jth soils and
x_{ki} is the value of the kth variate for the ith soil standardized by subtracting the mean for the 25 soils and dividing by the standard deviation for the 25 soils.

For soils 1 and 2, therefore, the generalized distance is given by:

$$d_{12} = \left| \left(\frac{15\cdot21 - 25\cdot50}{23\cdot26} - \frac{33\cdot27 - 25\cdot50}{23\cdot26} \right)^2 \right.$$

$$+ \left(\frac{70\cdot6 - 191\cdot48}{321\cdot37} - \frac{67\cdot5 - 191\cdot48}{321\cdot37} \right)^2$$

$$+ \left(\frac{467\cdot1 - 608\cdot96}{653\cdot68} - \frac{1059\cdot8 - 608\cdot96}{653\cdot68} \right)^2$$

$$+ \left(\frac{1400 - 1247}{644\cdot44} - \frac{460 - 1247}{644\cdot44} \right)^2$$

$$+ \left(\frac{0\cdot12 - 0\cdot108}{0\cdot065} - \frac{0\cdot15 - 0\cdot108}{0\cdot065} \right)^2$$

$$+ \left(\frac{0\cdot63 - 0\cdot820}{0\cdot634} - \frac{1\cdot19 - 0\cdot820}{0\cdot634} \right)^2$$

$$\left. + \left(\frac{4\cdot53 - 4\cdot51}{0\cdot909} - \frac{4\cdot90 - 4\cdot51}{0\cdot909} \right)^2 \right|^{1/2}$$

$$= (0\cdot602859 + 0\cdot000093 + 0\cdot822128 + 2\cdot127604 + 0\cdot213018$$
$$+ 0\cdot780185 + 0\cdot165682)^{1/2}$$
$$= (4\cdot711569)^{1/2} = 2\cdot17$$

As this calculation has to be repeated for all possible pairs of soils, i.e. for $n(n-1)/2$ pairs (and, in this case, for 300 pairs) it is clearly a job for a computer!

From the half matrix of distances between every pair of soils, the minimum spanning tree can be calculated, again using a computer, by one of several algorithms, but one of the most convenient is that given by Gower and Ross.[29] The results are shown in Table 6.12 and diagrammatically in Fig. 6.4. Many of the soils show marked similarities, but a few, and notably soil no. 3, show fairly distinct differences.

The minimum spanning tree is a valuable tool in its own right, as well as in helping with the interpretation of methods of cluster analysis by enabling the adequacy of clusters to be judged from the extent to which close neighbours are assigned to different clusters. One particularly important use is in vector diagrams that illustrate approximations in a few dimensions to configurations in many dimensions. The variability in this example is seven-dimensional, and any attempt to express that variability in fewer dimensions necessarily results in some distortion—the extent of that distortion can be judged by super-

Table 6.12   Minimum spanning tree for soils data.

Soil number	Joined to soil number	Distance
2	17	2·09
3	4	6·83
4	23	1·93
5	16	2·62
6	20	0·96
7	9	0·85
8	10	0·68
9	1	0·80
10	1	0·65
11	8	0·82
12	2	2·44
13	22	1·84
14	8	0·92
15	14	1·27
16	18	1·35
17	7	0·55
18	7	0·52
19	21	0·72
20	10	0·75
21	20	0·41
22	18	0·51
23	8	0·62
24	25	1·84
25	12	1·11

imposing the minimum spanning tree on the representation of the variability. Figure 6.5, for example, shows the distribution of the 25 soils on the two-dimensional plane represented by the variables of extractable iron and pH. The diagram clearly shows the anomalous position of soil no. 3, which might otherwise be assumed to be similar to soils 1, 9, and 19. Figure 6.6, which shows the distribution of the 25 soils on the two-dimensional plane of loss on ignition and pH, is, perhaps, even more revealing. Without the minimum spanning tree, it would be tempting to assume that soils 2 and 4 were similar, whereas soil 2 is more generally similar to soils 7, 11 and 17, while soil 4 is closer in general similarity to soil 23.

The single linkage cluster analysis obtained from the minimum spanning tree by clustering at threshold distances of 0·75, 1·00, 1·25, 1·50, etc. is given in Fig. 6.7. The analysis indicates the presence of several tight clusters (e.g. soils 1, 8, 10, and 23, soils 7, 17, 18 and 22, and soils 19, 20 and 21) linked by individual soils to form a major group of Lake District soils to which soil 3, and to a lesser extent soil 5,

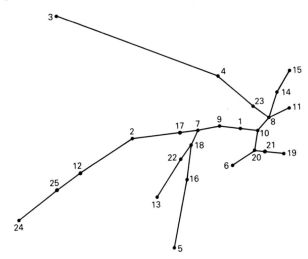

Fig. 6.4 Diagrammatic representation of minimum spanning tree for Lake District soils.

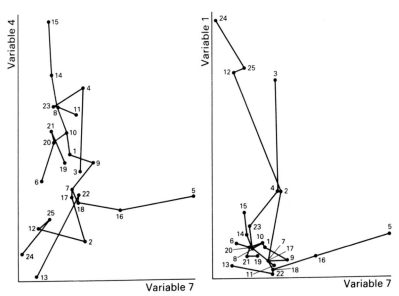

Fig. 6.5 Minimum spanning tree plotted on the plane of variability represented by extractable iron and pH.

Fig. 6.6 Minimum spanning tree plotted on the plane of variability represented by loss on ignition and pH.

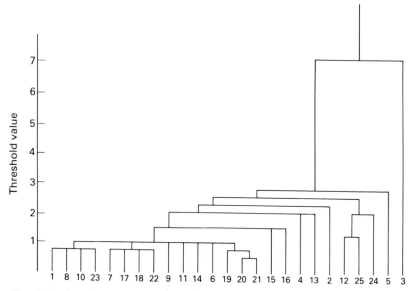

Fig. 6.7   Dendrogram for cluster analysis of Lake District soils.

is an outlier. The way in which the soils should be chosen for experiments on the response of sycamore and birch to soil nutrition will depend upon the precise objectives of the study. If a fairly homogeneous group of soils is required, only those soils linked at a low threshold distance should be included (i.e. soils 1, 8, 10, 23, 7, 17, 18, 22, 9, 11, 14, 6, 19, 20 and 21). If a full range of soil variability is desirable, only a few of these more homogeneous soils should be included, together with the more outlying soils, and, of course, soil 3.

Reciprocal averaging

Where some of the variates in descriptive models are qualitative, principal component analysis is of less value. Cluster analysis can still often be used, even on wholly qualitative data, by devising indices of similarity which are then translated into distances by formulae such as

$$d_{pq}^2 = 1 - S_{pq}$$

where the similarity S_{pq} between individuals is the ratio of 'matches' to the number of variates compared, a 'match' occurring when the variate is either present or absent in both individuals.

There are, however, some specially devised multivariate models for

qualitative data, and one of the most valuable of these is that of reciprocal averaging, described by Hill.[34] The model is especially appropriate for presence-absence data which are commonly found in ecology,
as in the recording of the presence or absence of particular species in
quadrats. Geometrically, these data can be regarded as a set of points
situated at the vertices of a hypercube, for which the ordination does
not depend on the explicit use of the distances between the vertices.

The method uses a scheme of successive approximations during
which the individuals are given an arbitrarily chosen set of starting
scores, ideally chosen to represent some gradient suspected *a priori* as
being reflected by the data. Average scores are then computed for each
variate from which new, rescaled averages are calculated for the
individuals. After a sufficient number of iterations, the final variate
scores converge to the same row vector, and the eigenvalue of the first
axis is a measure of how much the range of the scores contracts in one
iteration.

When the first axis has been obtained, the second is considered, and
a good starting point for the scores of the second axis may be obtained
by using a set of scores which were close to the final ones for the first
axis. Before iteration, however, these scores have to be adjusted by
subtracting a multiple of the final first axis. A simple example of the
computations, which are laborious for any practical application and
should be entrusted to a computer, is given by Hill.[34]

The process is essentially a repeated cross-calibration which derives
a unique one-dimensional ordination of both the variates and the
individuals. It is called 'reciprocal averaging' because the variate scores
are averages of the individual scores, and, reciprocally, the individual
scores are averages of the variate scores. The final scores do not depend
on the initial scores, but a good choice of initial scores considerably
reduces the number of iterations required. The whole procedure is
mathematically very similar to principal component analysis, and can
be extended to cover quantitative as well as qualitative data.

In the ecological applications, reciprocal averaging is usually
extended to what has come to be known as 'indicator species analysis'.
In this analysis, the individual quadrats are ordered by the first axis of a
reciprocal averaging ordination, and the quadrats are then divided into
two groups at the centre of gravity of the ordination. Five 'indicator
species' are chosen by the function

$$I_j = |m_1/M_1 - m_2/M_2|$$

where I_j is the indicator value of species j (assuming the value 1 if
        the species is a perfect indicator and a value 0 if it has no
        indicator value)

m_1 is the number of quadrats in which species j occurs on the negative side of the dichotomy

m_2 is the number of quadrats in which species j occurs on the positive side of the dichotomy

M_1 is the total number of quadrats on the negative side of the dichotomy

M_2 is the total number of quadrats on the positive side of the dichotomy.

The five species with the highest indicator values are then used to construct an 'indicator score' for the whole set of quadrats and to define an 'indicator threshold' which corresponds with dichotomy.

The whole process may be repeated for the second and subsequent reciprocal averaging axis, so that the quadrats are again divided, using the same method, and continuing as far as possible. No satisfactory rule for stopping the subdivision has so far been devised, so that there is some degree of arbitrariness both in the selection of thresholds and the number of subdivisions. The indicator scores can be regarded as providing an ordination of the quadrats on a six-point scale, admittedly a crude ordination, but one which can be done quickly, even in the field, by counting the presence of five selected species, in particular quadrats. The aim is to mirror the original reciprocal averaging ordination sufficiently closely for the analysis to be used as a criterion for classifying the quadrats.

Example 6.3 – Scottish native pinewoods.

Hill *et al.*[35] give an example of the application of reciprocal averaging and indicator species analysis to the classification of the native pinewoods in Scotland. The data for the analysis were derived from a survey of the 26 major native pinewoods described by Steven and Carlisle.[76] In the survey, sixteen 200 m^2 plots were selected at random within each of the cartographic units shown as pinewoods in maps by Steven and Carlisle. Each plot was subdivided into five superimposed quadrats of 4 m^2, 25 m^2, 50 m^2, 100 m^2 and 200 m^2.

A cumulative record of all vascular plants was assembled for these quadrats, working outwards from the centre. After the primary vegetation data had been collected, the trees and shrubs within the plot were measured, a list of standard habitat features was recorded, and major soil features noted from a shallow pit at the centre of the plot.

Only the presence-absence species data were used in the descriptive classification, the assumption being that each plot contained enough species for its floristic affinities to be judged from these data alone. The data actually consisted of records of 416 plots (26 sites × 16 plots) and included 176 species with a frequency greater than three.

Table 6.13   Classification of plot data for Scottish native pinewoods.

Division 1

Negative indicator species:	*Deschampsia flexuosa*
Positive indicator species:	*Drosera rotundifolia, Erica tetralix, Narthecium ossifragum, Sphagnum papillosum*

If indicator score $\leqslant 0$ go to division 2
If indicator score $\geqslant 0$ go to division 3

Division 2

Negative indicator species:	(none)
Positive indicator species:	*Agrostis canina, Anthoxanthum odoratum, Galium saxatile, Oxalis acetosella, Viola riviniana*

If indicator score $\leqslant 2$ assign to group AB
If indicator score $\geqslant 2$ assign to group CD

Division 3

Negative indicator species:	(none)
Positive indicator species:	*Agrostis canina, Galium saxatile, Luzula multiflora, Succisa pratensis, Viola riviniana*

If indicator score $\leqslant 2$ assign to group EF
If indicator score $\geqslant 2$ assign to group GH

The range of variation in the pinewoods was relatively limited and was well summarized by a two-dimensional ordination. The first axis of the ordination was related to the depth of the organic layers of the soil. The second axis was related to the acidity of the soil. Table 6.13 gives the first two levels of the hierarchy generated by the indicator species analysis of the plot data. The classificatory hierarchy is represented by the following scheme:

$$\{(AB)(CD)\}\{(EF)(GH)\}$$

indicating that the first division separates the groups A, B, C, and D from the rest, and that the second division divided the sub-group consisting of A, B, C, and D into two further groups A and B, and C and D, and so on. The relationship of the groups to the ordination is shown in Fig. 6.8 on which the means and standard deviations of the plot types are plotted.

The eight plot types derived from this descriptive model were found to be closely correlated with other environmental factors. The model therefore provides a framework for further detailed study rather than a solution to the dynamics of the pinewoods themselves. In particular, the classification, while easy to do in the field, enables a description and interpretation of the main site types to be made, and, while this

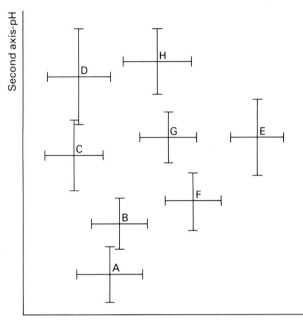

Fig. 6.8   Relationship of indicator species plot types to reciprocal averaging ordination for native Scottish pinewoods.

description broadly confirms the site groups defined by Steven and Carlisle,[76] it also suggests differences which warrant further investigation—one of the main purposes of any systems analysis.

PREDICTIVE MODELS

If we now turn to the broadly predictive models, the classification of Fig. 6.1 makes a distinction between models for which only one variate is predicted from two or more other variates, and those for which several variates are predicted from two or more other variates. Multiple regression analysis, which we have already encountered in Chapter 5 on stochastic models, is, of course, one type of predictive model which enables the value of one variate to be predicted from the values of two or more variables, usually known as regressor variables. Where the regressor variables are, in fact, variates, that is variables with a specified relative frequency or probability, it can be shown that, mathematically, the estimation procedures are equivalent even when there is a mixture of regressor variates and regressor variables in the one model. Thus, in practice, there is usually little to worry about with classical regression models for variates, variables, or a mixture of both, providing that they are reasonably error-free measurements of what they are supposed to measure.

When dealing with predictions of one variate, therefore, we shall concern ourselves with the classical model of discrimination in this chapter, for which the basic theory has been known for over forty years (Fisher[21]). As is indicated by Fig. 6.1, we shall distinguish between a model providing a discrimination between two groups, and known as the discriminant function, and a model providing a discrimination between more than two groups, and known as canonical variates.

Discriminant function

Fisher's classical model of the discriminant function deals with the problem of how best to discriminate between two *a priori* groups, each individual of which has been measured in respect of several variables. The model provides a linear function of the measurements on each variable such that an individual can be assigned to one or other of the two groups with the least chance of being misclassified. The discriminant function is written as:

$$z = a_1 x_1 + a_2 x_2 + \cdots + a_m x_m$$

where $\mathbf{a}$ is the vector of discriminant coefficients and $\mathbf{x}$ the vector of observations or measurements made on an individual which is to be assigned to one or other of the two groups. Note that, for this model, we are only considering the possibility of two groups, and, while we may decide that some individual cannot be assigned with any confidence to one of the two groups, we are not considering the formation of further groups. The model is again most simply explained through an example.

Example 6.4 – Vascular plants on Signy Island.

Signy Island belongs to the South Orkney Islands, situated within the maritime Antarctic region. The nearest land mass is the Antarctic peninsula, some 640 km away, but the nearest sources of immigrant organisms lie to the north-east at South Georgia, some 900 km away, and to the north-west at Tierra del Fuego, some 1440 km away.

In a study of the vegetation of the island, an arbitrary grid of 500 m^2 squares was imposed on the 1 : 25 000 scale map of Signy Island. Within these squares, environmental variables were assessed from maps prepared by scientific expeditions to the island. Similarly, the areas

Table 6.14   Summary of environmental variables.

Variables	Minimum	Mean	Maximum	Standard deviation
1. Maximum altitude (metres)	5	140·0	280	79·3
2. Number of contours cut by East–West transect	0	7·5	22	5·44
3. Percentage slope facing south	0	19·2	100	25·0
4. Percentage occupied by lakes	0	1·2	20	3·47
5. Percentage shown as rock	0	13·3	45	9·12
6. Percentage shown as drift and scree	0	27·2	91	25·8
7. Distance of sea to E (metres)	0	1026	4100	1084

within each square occupied by various types of vegetation were assessed from hitherto unpublished maps, and, in particular, the presence or absence of vascular plants in the square was recorded.

Preliminary analysis of the environmental variables suggested that the variables describe a total variability which is approximately seven-dimensional, and that the environmental variables can therefore be reduced to seven without significant loss of information. The values of these seven variables for 104 grid squares are summarized in Table 6.14. The presence of vascular plants was recorded for 22 of the squares, and the question of interest is whether the seven environmental variables, or some sub-set of these seven, can be used to predict the presence or absence of vascular plants.

The same variables have been measured for both groups of squares, so that the basic data may be arranged in two matrices, the first of order $n_1 \times m$ and the second $n_2 \times m$. As vascular plants were found in 22 of the squares:

$$n_1 = 22, \qquad n_2 = 82, \qquad m = 7$$

The upper triangle of the symmetric pooled variance–covariance matrix for the two data matrices is given in Table 6.15. The vector of discriminant coefficients **a** is obtained by the solution of simultaneous equations expressed

$$\mathbf{Sa} = \mathbf{d}$$

where **S** is the pooled variance–covariance matrix and **d** the difference

Table 6.15   Pooled variance–covariance matrix for data sets.

1	2	3	4	5	6	7
5310·961	199·8022	194·8896	−32·94015	−16·05446	62·73615	12569·94
	26·88293	−25·31391	1·559248	12·8375	42·83095	4·190566
		578·6623	−4·799063	−36·08125	−116·5440	−315·8835
			11·96896	3·166493	15·54125	62·82278
				83·73383	44·49558	1250·437
					667·5128	2087·470
						1175·907

between the mean vectors of the two groups. This equation is conveniently solved by premultiplying both sides by the inverse of the covariance matrix, obtaining

$$\mathbf{a} = \mathbf{S}^{-1}\mathbf{d}$$

The inverse of the pooled variance–covariance matrix is given in Table 6.16, the exponential notation being used for convenience, and the matrix again being symmetric about the principal diagonal. The mean vectors, the vector of differences, and the vector of discriminant coefficients are given in Table 6.17.

Hotelling's T^2 is then computed from the relation

$$T^2 = \frac{n_1 n_2}{(n_1 + n_2)}\,\mathbf{d}'\mathbf{a} = 35\cdot90$$

and the significance of T^2 assessed by the variance ratio test given by:

$$F = \frac{n_1 + n_2 - m - 1}{(n_1 + n_2 - 2)m}\cdot T^2 = 4\cdot83$$

with degrees of freedom $m = 7$ and $(n_1 + n_2 - m - 1) = 96$. If both groups had been taken from the same population, the probability of obtaining a value of F as high as this is approximately 0·00011.

The function:

$$Z = -0\cdot00880x_1 - 0\cdot142x_2 - 0\cdot0303x_3 - 0\cdot131x_4 + 0\cdot0226x_5$$
$$+ 0\cdot0120x_6 - 0\cdot000169x_7$$

therefore, provides a significant discrimination between 500 m² with and without vascular plants. The values of Z can be computed for each individual square in the two groups, the centroids of the discriminant scores for the two groups being −0·959 and −3·029 for the squares with and without vascular plants respectively.

The difference between the group centroids is known as the 'generalized distance' between the two groups, and can be calculated

Table 6.16 Inverse of the pooled variance–covariance matrix.

3·236548E−4						
−2·98233E−3	0·07199744					
−1·895112E−4	2·899737E−3	1·965440E−3				
9·9986822E−4	−0·01020711	−4·002193E−4	0·0897032			
4·140633E−4	−9·022103E−3	2·661735E−4	−8·298558E−4	0·01397195		
8·999152E−5	−3·144589E−3	1·602335E−4	−1·504498E−3	−2·672139E−4	1·783831E−3	
−4·153426E−6	4·812380E−6	1·997326E−6	−1·198577E−5	−1·866131E−5	−3·709850E−6	9·222408E−7

Table 6.17 Vectors of means, differences and discriminant coefficients.

Variable	Mean of squares with vascular plants	Mean of squares without vascular plants	Differences	Discriminant functions
1	78·1818	123·5096	−45·3278	−8·799238E−3
2	4·2273	6·6538	−2·4265	−0·1421932
3	5·6818	18·0385	−12·3567	−0·0301576
4	0·3182	1·1250	−0·8068	−0·1312235
5	14·2273	10·2596	3·9677	0·0226488
6	30·0000	20·8558	9·1442	0·0104919
7	825·0000	851·9232	−26·9232	−1·693091E−4

from T^2 by the relationship

$$D^2 = \frac{n_1 + n_2}{n_1 n_2} \cdot T^2$$

or more directly from

$$D^2 = \mathbf{d}'\mathbf{S}^{-1}\mathbf{d} = 2 \cdot 070.$$

A measure of the effectiveness with which the calculated discriminant function can be used to classify further individuals is provided by the standardized normal deviate

$$D/2 = 0 \cdot 719$$

Tabulated values of the cumulative normal distribution corresponding to this deviate give about 76 per cent of the individuals as correctly assigned to their group by the discriminant function.

The calculation of the discriminant scores allows us to view the individual squares originally assessed as two clusters of points, each centred on its mean scores, the points being dispersed along a line. The value of the discriminant function lies partly in determining the relevance of the basic variables to the discrimination and partly in the ability to compute a discriminant score for any new square to be assigned to one of the two groups—or perhaps to be set aside as not belonging to either. We may wish to decide, for example, whether or not some arbitrary 500 m² square superimposed on the map of Signy Island is likely to contain vascular plants in order to determine whether a search for those plants should be attempted under relatively difficult conditions in the field.

If the two groups from which the discriminant coefficients were calculated are equally represented in the population from which the samples were drawn, the mid-point between the group centroids is an appropriate division for the assignment of an unclassified individual to one of the two groups. In this example, therefore, squares with a discriminant score less than $-1 \cdot 994$ would be allocated to the group with no vascular plants, while squares with a discriminant score greater than $-1 \cdot 994$ would be allocated to the group with vascular plants.

Where the two groups are unequally represented in the population, the point of division must be shifted from the midpoint between the two centroids toward the smaller group by a distance equal to

$$\frac{\log_e R}{D}$$

where R is the ratio of the number of individuals in the larger group to the number of individuals in the smaller group. If we can assume that the squares included in the analysis are an unbiased sample from the population of possible squares, then

$$R = 3\cdot727$$

and

$$\frac{\log_e R}{D} = \frac{1\cdot316}{1\cdot439} = 0\cdot914$$

so that the point of division should be shifted from $-1\cdot994$ to $-1\cdot080$.

Canonical variates

The discriminant function provides a powerful and practical model for discriminating between two *a priori* groups. There are many ecological situations, however, in which we wish to discriminate between more than two groups. In the simplest of such situations, a third group may be formed by the hybrids between the two original groups, as in the study of hybridization between Red deer and Sika deer by Lowe and Gardiner.[52] Even in taxonomic research, however, considerably more complex problems may be encountered as in the studies of grasshoppers by Blackith and Blackith[6] and in the studies of the ecology of insects living on broom plants by Waloff.[89]

As for any other multivariate model, we again have a matrix of observations with the columns representing p variates for each of n individuals in the rows of the matrix. In contrast to the model for principal components, an *a priori* structure is imposed on this matrix, the individuals being derived from m ($<n$) separate groups. While we will again seek to represent the p variables in as few dimensions as possible, we will also wish to emphasize between-group variability at the expense of within-group variability. The necessary distinction is between a matrix of the pooled within-group sums of squares and products of deviations from the group means, **W**, and a matrix of between-groups sums of squares and products of the deviations of the group means from the overall means, **B**.

The object is now to derive not one discriminant function, but a set of such functions of the form

$$d = a_1 x_1 + a_2 x_2 + a_3 x_3 + \cdots + a_p x_p$$

where $a_1, a_2, a_3, \ldots, a_p$ are discriminant coefficients computed so as to minimize the confusion between one group and another. It is relatively easy to show (see, for example, Krzanowski[44]) that these coefficients are provided by the eigenvalues and eigenvectors of the product of the two matrices $\mathbf{W}^{-1}\mathbf{B}$. The elements of the normalized eigenvectors are the weights $a_1, a_2, a_3, \ldots, a_p$ while the eigenvalues are measures of the discriminatory power associated with each of the canonical variates or discriminant axes.

Example 6.5 – Leaf shape of poplar clones.

As part of a programme of research on the behaviour and characteristics of poplar varieties, collections of leaves were made from many varieties and a series of standard measurements recorded for each leaf. Nine of the measurements were subsequently selected for further investigation, as follows:

1. Length of petiole
2. Length of leaf blade
3. Width of leaf blade at its widest point

Table 6.18　Basic leaf measurements of poplar.

Variety	1	2	3	4	5	6	7	8	9
Populus serotina VB	39	98	95	55	88	76	110	90	100
	39	96	95	58	92	80	95	90	92
	43	100	84	54	80	70	100	70	97
	54	123	117	72	113	94	105	85	110
	47	114	105	69	102	90	92	72	95
Populus gelrica HA	47	108	121	95	118	110	110	87	110
	56	90	120	95	117	110	105	80	100
	65	130	140	95	140	125	115	90	110
	50	114	118	85	113	108	110	85	108
	47	113	125	87	121	110	110	90	108
Populus gelrica VB	60	120	132	90	122	114	115	85	110
	44	87	101	67	97	88	100	75	97
	45	94	100	66	88	86	105	85	108
	59	115	125	84	118	106	110	70	110
	49	90	107	75	103	96	110	90	108
Populus T × T32	24	117	84	60	84	76	80	60	90
	30	134	105	73	104	92	65	50	87
	31	150	114	73	110	96	70	60	85
	23	140	90	64	95	87	60	50	85
	27	126	98	75	96	90	77	60	92
Populus T × T37	12	118	61	43	59	52	60	50	70
	15	136	95	55	89	75	55	65	90
	16	145	101	63	97	84	90	60	90
	17	161	118	64	112	94	85	65	90
	18	155	105	60	100	83	70	70	95
Populus serotina erecta	58	130	124	64	114	84	90	50	90
	53	124	118	60	110	94	95	55	90
	67	134	130	80	126	103	95	50	85
	57	124	120	72	114	100	93	65	92
	56	122	124	70	120	100	95	65	93

Table 6.19 Half-matrix of between-group sums of squares and products.

Variable		1	2	3	4	5	6	7	8	9
P	1	6992·73								
L	2	−5031·18	6896·12							
w_1	3	4616·37	−2147·25	4244·50						
w_2	4	3112·37	−2580·44	3070·87	3683·00					
w_3	5	4119·93	−1765·49	3940·25	3115·75	3790·87				
w_4	6	3826·18	−2572·25	3676·00	3744·56	3615·62	3993·31			
b	7	6237·19	−6589·62	3909·50	3477·56	3375·12	3792·75	7593·50		
v_1	8	2616·40	−4863·69	1308·68	2255·43	1126·75	2010·06	5035·19	4732·56	
v_2	9	2467·59	−3483·37	1560·87	2064·62	1387·75	2006·37	3708·37	2980·56	2147·62

Table 6.20 Half-matrix of within-group sums of squares and products.

Variable	1	2	3	4	5	6	7	8	9
1	808·40								
2	1227·40	4189·59							
3	1332·39	3664·59	4251·19						
4	361·80	1661·20	2128·61	1522·40					
5	1344·79	3533·62	4069·18	2123·40	4046·81				
6	1013·00	2677·81	3116·20	1765·59	3110·79	2614·00			
7	266·61	832·42	1103·59	556·20	997·02	867·00	1615·61		
8	−118·80	207·59	611·00	138·00	513·41	518·80	503·20	1298·40	
9	254·20	1022·01	1131·57	503·98	1013·18	813·61	597·81	551·00	835·61

Table 6.21 Discriminant matrix for poplar varieties.

Variable	1	2	3	4	5	6	7	8	9
1	680·664	-570·221	386·839	236·172	326·647	300·492	649·240	307·849	259·044
2	-347·187	293·287	-217·448	-159·342	-188·698	-189·260	-338·606	-165·818	-147·135
3	301·905	-179·680	201·352	72·422	157·388	120·518	244·195	53·144	76·830
4	-490·526	332·712	-277·978	-105·947	-224·159	-170·362	-403·467	-122·630	-128·989
5	-263·822	202·724	-179·153	-111·376	-144·730	-142·457	-242·703	-89·089	-103·355
6	361·501	-286·830	242·426	191·883	214·392	223·842	327·599	137·036	147·628
7	54·913	-54·648	29·868	17·315	23·828	21·749	64·801	41·007	27·250
8	-40·764	-15·374	-39·973	-12·537	-35·114	-24·556	6·423	47·899	12·943
9	130·949	-163·570	74·260	92·256	66·186	92·650	160·506	110·427	90·137

Table 6.22 Eigenvalues of determinant matrix.

Canonical variate	Eigenvalue	Percentage	Cumulative percentage
I	1097·34	81·21	81·21
II	168·50	12·47	93·68
III	58·42	4·32	98·00
IV	16·39	1·21	99·21

4. Width of leaf blade half-way along the blade
5. Width of leaf blade one-third along the blade
6. Width of leaf blade two-thirds along the blade
7. Distance of the base of the leaf from the point at which the petiole joins the blade
8. Angle of the first major vein with the mid-rib
9. Angle of the first minor vein with the mid-rib

All the linear measurements were recorded to the nearest millimetre. The two angles were measured to the nearest degree.

Table 6.18 gives the basic leaf measurements for five leaves from separate trees of each of six varieties of poplar. The varieties, therefore, reflect six *a priori* groupings of the leaves, for which we seek effective discriminators based on the nine measurements.

The half matrix of between-group sums of squares and products is given in Table 6.19. (Note that, because the matrix is symmetric, we need give only the principal diagonal—representing the between-group sums of squares—and either the upper or lower off-diagonal elements —representing the sums of products.) Similarly, the half-matrix of pooled within-group sums of squares and products is given in Table 6.20. The inverse of this symmetric matrix post-multiplied by the between-groups matrix gives the so-called determinant matrix shown in Table 6.21. Note that this determinant matrix is not itself symmetric.

The first four eigenvalues of the determinant matrix are summarized in Table 6.22, the percentage of the variability being calculated from the sum of the principal diagonal of the matrix which is equal to 1351·30. The first three canonical variates account for 98 per cent of the total variability of the determinant matrix, and the first of these variates accounts for 81·21 per cent. An appropriate test of significance suggests that the first three variates are of practical value in discriminating between the groups.

The weights given to the nine variables are summarized in Table 6.23 by standardized eigenvectors corresponding to the first four eigenvalues of the determinant matrix. The first canonical variate gives greatest weight to petiole length and to a general contrast of leaf shape, which gives positive weight to the width of the leaf blade at its widest point and two-thirds of the distance along the blade, and negative weight to leaf length and the width of the blade half-way and one-third of the way along the blade. Leaves with high values of this canonical variate tend therefore to have long petioles and short, broad, rapidly-tapering leaves, while leaves with a low value of the variate will tend to have short petioles and long, narrow, slowly-tapering leaves.

The remaining canonical variates are all variations of the width of the

Table 6.23 Weighting of variables in canonical variates.

| Variable | Standardized weights for canonical variates | | | |
	I	II	III	IV
1. Length of petiole	1·000	0·428	0·189	0·158
2. Length of blade	−0·507	0·011	0·287	0·089
3. Blade width (w_1)	0·423	0·716	−0·448	−0·829
4. Blade width (w_2)	−0·697	−1·000	0·107	0·209
5. Blade width (w_3)	−0·384	−0·183	0·631	1·000
6. Blade width (w_4)	0·513	−0·059	−1·000	−0·357
7. Leaf base	0·087	0·003	0·196	0·026
8. Major vein angle	−0·035	−0·255	0·622	0·198
9. Minor vein angle	0·211	−0·384	−0·118	−0·271

leaf blade. The second canonical variate is largely a contrast between the width of the blade half-way along the blade and its width at the widest point of the blade. The third canonical variate is a contrast between the width of the blade two-thirds along the blade, and one-third along the blade, together with the major vein angle. The fourth canonical variate is a contrast between the width one-third of the distance along the blade with the width of the blade at its widest point. These three measures of taper, however, add only 18 per cent of further discrimination to the 81 per cent provided by the first canonical variate.

Finally, the mean values of the four canonical variates for each of the six groups are given in Table 6.24. The two T × T crosses are immediately identifiable from their low values of the first canonical variate, but are not otherwise easily discriminated by these data. *Populus serotina erecta* is readily discriminated from the remaining three varieties by its relatively high value of the second canonical variate, while the two varieties of *Populus gelrica* are identifiable from each other by the fourth canonical variate and both from *Populus serotina* by the third canonical variate. As before, the minimum spanning tree (this time based on the canonical variates) helps in the interpretation of the affinities of the six varieties—see Fig. 6.9.

The multivariate model of canonical variate analysis is a powerful and extremely flexible tool for the investigation of the ability to discriminate between several *a priori* groups, and is a logical extension of the discriminant function.

Canonical correlations

Finally, we come to the multivariate model appropriate to one of the most difficult of all statistical problems—defining the relationships between two or more sets of variates. We have already suggested that

Table 6.24 Mean values of canonical variates for poplar varieties.

Variety	Mean values of canonical variates			
	I	II	III	IV
Populus serotina VB	5·51	−10·18	8·18	0·80
Populus gelrica HA	6·53	−13·95	4·94	1·05
Populus gelrica VB	8·40	−11·72	5·21	−1·11
Populus T × T32	−4·22	−11·28	4·05	0·68
Populus T × T37	−7·25	−10·04	6·43	−0·22
Populus serotina erecta	6·90	−6·10	4·43	0·58

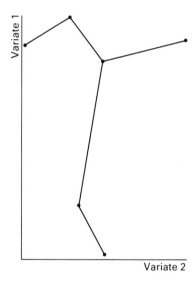

Fig. 6.9 Minimum spanning tree for poplar varieties plotted on first and second canonical variates.

one possible approach to this problem is through the model of principal component analysis. By calculating the principal components of each set of variates and then correlating the resulting components, we may frequently deduce the complex interrelationships between these sets. A study using this approach to investigate the relationship between growth of Corsican pine in southern Britain and the physiography, soil chemistry and soil physics of the sites on which the trees were growing is described by Fourt *et al.*[25]

As an alternative to the assumption of an *a priori* structure imposed

Table 6.25 Coefficients of correlations between species and environmental variables.

Species	Coefficients of correlations with environmental variables							
	>250 μm	125–250 μm	62·5–125 μm	<62·5 μm	Loss on ignition	Calcium	Phosphorus	Nitrogen
Macoma balthica	−0·040	−0·239‡	0·159‡	0·113	0·402‡	0·158†	−0·023	0·516‡
Tellina tenuis	−0·016	0·022	0·055	−0·077	−0·016	−0·035	−0·006	−0·063
Hydrobia ulvae	−0·068	−0·232‡	0·201‡	0·071	0·171†	0·038	−0·022	0·194†
Corophium volutator	−0·064	−0·265‡	0·087	0·224‡	0·092	0·060	−0·024	0·133*
Nereis diversicolor	−0·011	−0·152*	0·056	0·114	0·299‡	0·106	−0·006	0·401‡
Arenicola marina	0·244	−0·048	−0·027	0·032	0·096	0·230‡	0·083	0·139*
Nephthys hombergii	−0·050	0·095	0·113	−0·209‡	−0·068	−0·090	−0·020	−0·110

For details of variables, see Example 6.1, p. 103 *et seq.*
* significant at 0·05.   † significant at 0·01.   ‡ significant at 0·001.

on the individuals, or rows, of the basic data matrix, we now assume that the variates, or columns, of the matrix can be divided into two sets, with r and q variates in each set, so that $p = r + q$. This is equivalent to writing the data matrix as

$$\mathbf{X} = (\mathbf{x}_1 \mathbf{x}_2)$$

where $\mathbf{X}_1$ has n rows and r columns and $\mathbf{X}_2$ has n rows and q columns. The covariance or correlation matrix computed from the basic data matrix can therefore be partitioned as

$$\mathbf{S} = \begin{bmatrix} \mathbf{A}_{rr} & \mathbf{C}_{rq} \\ \mathbf{C}'_{qr} & \mathbf{B}_{qq} \end{bmatrix}$$

From this matrix, it will be of interest to find the linear combinations:

$$u_i = l'_i \mathbf{X}_1$$
$$v_i = m'_i \mathbf{X}_2$$

where $i = 1, 2, 3, \ldots, s$ with the property that the correlation of u_1 and v_1 is greatest, the correlation of u_2 and v_2 is greatest among all linear combinations uncorrelated with u_1 and v_1, and so on for all s possible pairs.

It is possible to show mathematically that these canonical correlations and the vectors describing the weighting necessary for each variable of the two sets are given by the eigenvalues and eigenvectors of the two matrices:

$$\mathbf{C}'\mathbf{A}^{-1}\mathbf{C} - \mathbf{B}$$
$$\mathbf{C}\mathbf{B}^{-1}\mathbf{C}' - \mathbf{A}$$

Both matrices have identical eigenvalues and their eigenvectors give the coefficients for the linear combinations of the left-hand and right-hand variables respectively.

Example 6.6 – Physical environment and invertebrates in Morecambe Bay.

In Example 6.1, we examined the correlations between the principal components of two sets of variables, one representing the physical and chemical properties of the sands and muds of Morecambe Bay, and the other representing the numbers of seven species of invertebrate organisms found in the sand and mud samples. Clearly, this is the type of situation in which the alternative model of canonical correlation might be expected to be useful.

Table 6.25 gives the correlations between the eight environmental variables and the seven species variables for the 272 samples for which

Table 6.26   Eigenvalues and canonical correlations.

Number	Eigenvalue	Canonical correlations	Proportion of variance Species	Proportion of variance Environment
1	0·312523	0·559	0·25	0·24
2	0·111680	0·334	0·13	0·18
3	0·080338	0·283	0·16	0·19
4	0·044555	0·211	0·13	0·13

both sets of variables were available. This table represents the matrix C_{rq} in the notation used above, with seven rows of species (r) and eight columns of physical and chemical variables (q). From this matrix and those of A_{rr} and B_{qq} which have values close to those of Tables 6.6 and 6.2 respectively—remember that there were only 272 samples for which both sets of variables were available—the eigenvalues of Table 6.26 can be calculated. Marriott[54] gives a simple test of the significance of these correlations, and, by this test, only the first two of these correlations is significant, although the third is close to significance. As a very rough 'rule of thumb', canonical correlations less than 0·3 are unlikely to be shown as significant, but canonical correlations just below this value are usually worth examining to see if they have some meaning that can be interpreted.

The first three canonical correlations account for 54 per cent of the variation in species and 61 per cent of the variation in the environmental variables. In contrast to the correlations of the components of the two sets of variables in Example 6.1, the canonical correlation analysis gives a direct measure of how much of the variability of the two sets can be accounted for by the relationships between them—in this example the degree of interrelationship is surprisingly high.

The scaled vectors for the species and environmental variables are summarized in Tables 6.27 and 6.28 respectively. The first of the correlations gives the highest weight to the numbers of *Macoma*

Table 6.27   Scaled vectors for species variables.

Species	Scaled vectors for correlation 1	2	3
Y_1 *Macoma balthica*	1·000	−0·326	−0·230
Y_2 *Tellina tenuis*	−0·067	−0·431	0·016
Y_3 *Hydrobia ulvae*	0·012	−0·474	−0·674
Y_4 *Corophium volutator*	0·146	1·000	−1·000
Y_5 *Nereis diversicolor*	0·194	0·007	0·725
Y_6 *Arenicola marina*	0·067	0·786	0·964
Y_7 *Nephthys hombergii*	0·007	−0·910	−0·099

Table 6.28   Scaled vectors for environmental variables.

Environmental variables	Scaled vectors for correlation		
	1	2	3
X_1 Per cent of particles >250 μm	0·060	0·253	0·104
X_2 Per cent of particles 125–250 μm	0·913	0·765	−0·616
X_3 Per cent of particles 62·5–125 μm	1·000	0·720	−0·997
X_4 Per cent of particles <62·5 μm	0·666	1·000	−1·000
X_5 Per cent loss on ignition at 550°C	0·015	−0·126	0·124
X_6 Per cent calcium	0·052	0·105	0·039
X_7 Per cent phosphorus	−0·005	0·062	0·174
X_8 Per cent nitrogen	0·936	−0·023	0·160

balthica amongst the species variables and to the percentage of particles between 62·5 and 250 μm and the percentage of nitrogen. The second of the correlations contrasts the numbers of *Corophium volutator* and *Arenicola marina* with the numbers of *Nephthys hombergii*, and, to a smaller extent, those of *Macoma balthica*, *Tellina tenuis* and *Hydrobia ulvae*. This contrast is correlated mainly with the percentage of fine particles <62·5 μm but also with the percentage of particles between 62·5 and 125 μm. Finally, the third correlation compares the contrast between the number of *Corophium volutator* and *Arenicola marina* with the absence of particles less than 125 μm.

The multivariate model of canonical correlations provides a powerful technique for summarizing and exploring complex relationships between two sets of variables. It is a model which has been neglected, mainly because of difficulties in computation. Efficient algorithms for these computations are now readily available, however, so that there is little excuse for continuing neglect.

This review of multivariate models has necessarily been very sketchy, and much more could be said about this potentially valuable class of models which make a direct attack upon the essentially multivariate nature of many ecological problems. Readers wishing to know more about these models will find further examples given by Seal,[70] Harris[32] and Kendall.[41]

The models exploit the mathematical properties of matrices of observations, and take account of the inherent structure and partitioning of those matrices. As in some of the other mathematical models, we exchange the freedom of dynamic simulation for the knowledge of the properties and behaviour of certain kinds of mathematics. The constraints are balanced by a surer insight into the logic of the model as an approximation to reality.

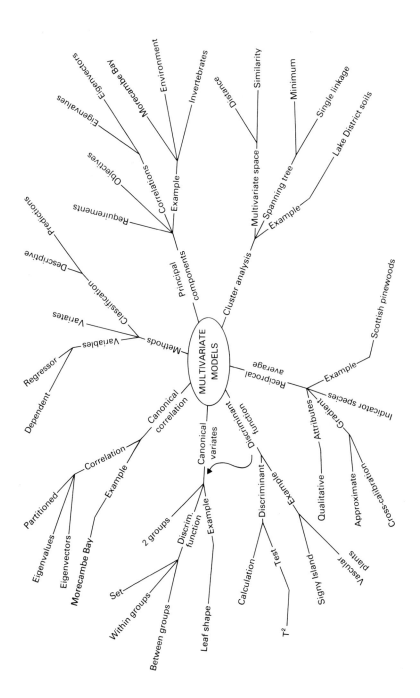

MULTIVARIATE MODELS

Principal components
- Requirements
- Correlations
  - Objectives
  - Example
    - Eigenvalues
    - Eigenvectors
    - Morecambe Bay
    - Environment
    - Invertebrates

Cluster analysis
- Distance
- Similarity
- Multivariate space
  - Minimum
  - Spanning tree
    - Single linkage
    - Example
      - Lake District soils

Methods
- Classification
  - Descriptive
  - Predictive
- Variables
  - Variates
  - Regressor
  - Dependent

Canonical correlation
- Correlation
  - Partitioned
  - Eigenvalues
  - Eigenvectors
- Example
  - Morecambe Bay

Canonical variates
- 2 groups
- Discrim. function
  - Set
  - Within groups
  - Between groups
  - Example
    - Leaf shape

Discriminant function
- Discriminant
  - Example
  - Calculation
  - Test
    - T^2
  - Sign
- Island
- Vascular plants

Reciprocal average
- Example
  - Scottish pinewoods
- Indicator species
- Gradient
- Attributes
  - Qualitative
  - Approximate
  - Cross-calibration

7

Optimization and Other Models

This chapter is the last of those reviewing the main families of mathematical models which are likely to be used in the application of systems analysis to ecology. It will be a shorter chapter than the others because the models which it describes are generally regarded as being less appropriate to ecology or have been relatively recently developed and so have not yet reached their full potential. Nevertheless, it is important for the systems analyst to be aware of the possible potential of the widest range of models, and the models described in this chapter should be borne in mind during the phase in which solutions are being generated. If this book should ever reach a second edition, there is a high probability that it will be this chapter which will need to be expanded to include new developments and applications.

OPTIMIZATION MODELS

'Optimization' is a rather horrid word that has been conjured out of the English language to describe finding the maximum or minimum of some mathematical expression or function by setting values to certain variables which we are free to alter within defined limits. If we always wished to find a maximum, we could describe the process as maximization—a word which is marginally more acceptable. Alternatively, we might wish to find a minimum and describe the process as minimization. Mathematically, one of these alternatives can always be turned into the other, so that there is some logic in regarding both processes as the same, hence 'optimization'. The logic is good, if the grammar is appalling!

Of course, nearly all of the models which we have so far examined in this book can be used to search for either an optimum or a minimum. Whether or not this is a sensible use of the model will depend entirely upon the context of the problem, but there will be many occasions on which we will want to explore the possibility of being able to increase the output of some ecological system by modifying its environment or by altering the methods of management. It is, indeed, one of the main reasons for the use of models that we should be able to look at the consequences of such changes.

In our dynamic model of the growth of yeasts in mixed culture (see page 28), for example, we may well wish to find the best combination of initial amounts of the two species in order to determine the maximum production of yeast cells. Similarly, we may wish to find the ideal starting mixture of barley and oats for the combined crop of the two species to be a maximum. Having determined the basic parameters of the models by our experiments, we may then carry out further experiments on the models to determine these combinations.

In the matrix models, we may experiment with several different age structures and harvesting rates to determine the optimum value of some objective function, despite the fact that the matrix methods themselves define the steady states and harvesting rates for given starting conditions. The stochastic models, too, provide interesting opportunities for this type of experimentation. The interested reader might, for example, like to determine the number of days exposed, the moisture content, and temperature for which oxygen uptake is a maximum in leaf litter from the example on pages 87–89. Similarly, it may be interesting to investigate the effects of changes in the transition probabilities of the raised mire model on the length of time that any area remains in a particular state.

Optimization is perhaps more difficult to envisage in the multivariate models, and, in one sense at least, these models have already found the optimum conditions defining the relationships. In discrimination, for example, the discriminant functions represent the linear functions of the original variables which give the best (i.e. optimum) discrimination between the *a priori* groups. Similarly, the canonical correlations define the linear functions of the two sets of variates which have the highest correlations.

We may, however, wish to formulate our model so that the search for the optimum combination of the critical variables is facilitated, and the underlying mathematics of this formulation was developed quite independently in the early application of mathematical techniques to practical problems that has now come to be known as 'operational research', or, in North America, as 'operations research'. Even more confusingly, the general class of solutions was known as 'mathematical programming' before the word 'programming' was adopted as the description of writing instructions for computers. The fact that most mathematical programming is now closely associated with computers only adds to the confusion!

The simplest form of mathematical programming to describe is known as 'linear programming'. In this model, we begin with a linear objective function:

$$Y = a_1x_1 + a_2x_2 + \cdots + a_nx_n = \sum a_ix_i$$

and we wish to make this function a maximum or a minimum subject to one or more constraints which are also expressed as linear functions, although, initially, these constraints may be inequalities, e.g.

$$b_1x_1 + b_3x_3 \geqslant Z$$

Frequently, there are implicit constraints that the x_i cannot be negative.

Where there are only two variables, optimization problems of this kind can be solved by graphical methods rather simply. For more than two variables, the problem rapidly becomes more difficult, and the usual approach to the solution is by what is known as the 'Simplex' method. In essence, the inequalities in the constraints are first removed by introducing some additional 'slack' variables. Any feasible solution to the problem is then sought, and, once such a solution has been found, iterative attempts are made to 'improve' the solution, i.e. move it closer to the defined optimum of the objective function by making small changes in the values of the variables. This iterative procedure continues until no further improvement can be made.

One of the advantages of optimization models is that they always illuminate two important facets of the problem. The solution gives the values of the variables of the objective function necessary for that function to be either a maximum or a minimum, depending on how the problem was defined. However, the method also indicates the constraint which needs to be relaxed for the optimum value of the objective function to be improved. In this way, the experimenter can examine more carefully his definition of the problem, and, in particular, his estimates of the coefficients of the variables in the objective function and the nature of the constraint. If he finds that the estimates can be improved, or the constraint relaxed, he may be able to find an even better solution.

As before, a simple example is perhaps the easiest way of demonstrating the structure and solution of the model.

Example 7.1 – Optimum predator strategies.

Chaston[10] suggests a simple ecological problem which is capable of being formulated as a linear programming model. He postulates the existence of a predator that has a nest at a site A and that has two potential sources of food x_1 and x_2 which are located in areas B and C respectively. The times taken to travel to areas B and C and return with a single unit of food are estimated as being two minutes and three minutes respectively. However, at site B, the predator takes two minutes to capture a unit of x_1, while at site C it takes only one minute to capture a unit of x_2. The calorific value of one unit of x_1 is estimated as 25 Joules, and the calorific value of one unit of x_2 as 30 Joules.

If we now add the constraint that the predator cannot devote more than 120 minutes per day in travelling to and from the nest and either of the two sites, and that it cannot spend more than 80 minutes per day searching for prey, we have a classical linear programming problem. These constraints can be written as the inequalities

$$2x_1 + 3x_2 \leqslant 120$$

for travelling time, and

$$2x_1 + 1x_2 \leqslant 80$$

for searching, together with the implied constraints

$$x_1 \geqslant 0 \qquad x_2 \geqslant 0$$

as the predator cannot capture a negative number of either species. Subject to these constraints, we wish to maximize the objective function

$$Z = (25x_1 + 30x_2) \text{ Joules}$$

A graphical solution can be found for this particular problem rather easily by concentrating our attention on the constraint inequalities. The constraint on travelling time ensures that, if x_1 is zero, x_2 has a limiting value of 40 units. Similarly, if x_2 is zero, x_1 has a limit of 60 units. The possible limiting combinations of x_1 and x_2 can then be represented in Fig. 7.1 as a straight line joining the two points ($x_1 = 60$, $x_2 = 0$) and ($x_1 = 0$, $x_2 = 40$). The same reasoning can be applied to the constraint on searching time, which ensures that, if x_1 is zero, x_2 has a limit of 80 units, while, if x_2 is zero, x_1 has a limit of 40 units. The possible limiting combinations of x_1 and x_2 can then be represented in Fig. 7.1 as a straight line joining the two points ($x_1 = 40$, $x_2 = 0$) and ($x_1 = 0$, $x_2 = 80$).

All of the feasible solutions to the problem lie in the space defined by O, P, Q, R in Fig. 7.1, and the maximum of the objective function will

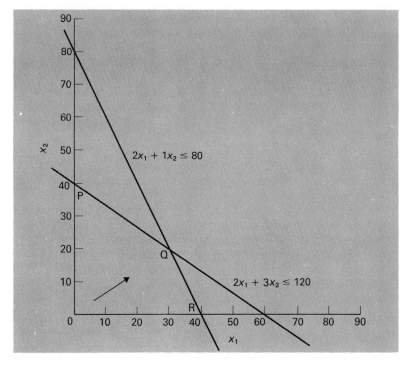

Fig. 7.1 Graphical solution of simple linear programming problem.

be the point which is furthest from the origin of the two axes in the direction of the arrow. This point coincides with the values of $x_1 = 30$ and $x_2 = 20$, so that the maximum value of the objective function is given by

$$Z = 25x_1 + 30x_2$$

$$= 25(30) + 30(20) = 1350 \text{ Joules}$$

The reader may like to test the effects of relaxing one or both constraints on this objective function, remembering that the greatest interest in optimization frequently lies in being able to identify the constraint which needs to be relaxed in order to find an even better solution.

This example is, of course, a very simple one, and the graphical method of solution can usually be employed only when there are two variables in the objective function and the constraints, although such solutions can often be found when there are several constraints.

However, well-defined algorithms exist for the solutions of linear programming problems, and particularly for those applications which can be formulated as transportation problems. Interested readers will find a simple introduction to linear programming given by Vajda.[85]

There are many examples of ecological problems which might be expressed as linear programming models. Most of these problems will be related to the management of natural resources in forestry and agriculture. Wardle[90] presents a linear programming study of forest management, while Fornstad[23] describes the linear programming planning system of the Swedish Forest Service. Olsson[59] describes a similar model for studying the growth of agricultural enterprises. One of the most interesting applications of the models is for the planning of research, and van Buijtenen and Saitta[86] describe some initial attempts to apply linear programming methods to the economic analysis of forest tree improvement. Russell[67] has developed an iterative system, based on linear programming, for selecting a portfolio of research projects such that the research outputs would provide society with the potential power to change the agricultural system in ways that are expected to bring about the greatest improvement in social welfare.

Useful though linear programming may be, it is easy to see that many problems will be difficult (or impossible) to express in terms of linear objective functions and constraints. Indeed, non-linearity in either the objective function or the constraints, or both, introduces quite disproportionate levels of difficulty in finding appropriate solutions. So, too, do problem formulations which impose limitations on the size of the lumps in which units of some particular resource can be allocated. There is, therefore, a well-developed theory of non-linear programming, although relatively few ecological models have drawn heavily upon this theory. Van Dyne et al.[87] have summarized a few of the important recent uses, and have introduced some potential needs, of optimization techniques in natural resource management.

Some large optimization problems can be reformulated as a series of smaller problems, arranged in sequences of time or space, or both. A reformulation of this kind is frequently desirable in order to reduce the computational effort of finding a solution, although care has to be taken to ensure that the sum of the optimal solutions of the sub-problems approaches the optimum solution of the whole problem. This search for the best solution at each of a number of stages is known as 'dynamic programming'. The mathematics of the models is frequently complex, and, for this reason, there are relatively few examples of successful examples in ecology. Hilborn[33] describes the use of stochastic dynamic programming for determining optimal harvesting rates for mixed stocks of fish. Watt[91] has suggested applications in determining strategies for insect pest control and Schreider[69] describes a model that covers the investment decisions for the whole production process in forestry, from tree seedling to the final product of one or more of the primary forest industries.

Interested readers who want to learn more about optimization will find the book by Converse[13] a useful introduction. In addition to broad study of the theory of optimization, the book contains some excellent general-purpose computer programs in BASIC.

GAME THEORY MODELS

Closely related to mathematical programming models are the models which are based on the theory of games. The simplest of these models to describe is known as the two-person zero-sum game. Such games are characterized by having two sets of interests represented, one of which may be Nature or some other external force, and by being 'closed', in the sense that what one player loses in the game the other must win. The theory can be extended to many-person non-zero-sum games, but the extension is well outside the scope of this intro-ductory text.

In the kind of game with which we are concerned here, the analysis centres on a simple matrix showing the strategies which can be adopted by the two players and the outcomes of all possible combina-tions of these strategies. As an illustration of this matrix, Table 7.1 shows the possible outcomes for a man who cannot remember whether or not a particular day is his wife's birthday! He has two possible strategies, i.e. he can buy some flowers or he can go home without any. Nature also has two strategies, i.e. it either is or is not his wife's birthday!

The figures in the main body of the table show the value of the game to the husband for each combination of the policies of the two players. Thus, if the husband goes home without flowers and it is not his wife's birthday, the value of the game to him is zero—he neither gains nor loses. If, on the other hand, he goes home without flowers and it is her birthday he loses heavily (− 10) for having forgotten something which he should have remembered. If he goes home with flowers and it isn't her birthday, there is a modest gain—perhaps blended with the suspicion that he has been doing something he shouldn't. If he goes home with flowers and it is her birthday, the gain is slightly greater (1·5) because he has remembered something he should.

It can be shown that this game has something that is called a 'saddle-point' in game theory. In simple terms, when the larger of the

Table 7.1 Game matrix showing strategies and outcomes for the 'wife's birthday?' problem.

	Strategies	Nature Not birthday	Birthday
Husband	Empty-handed	0	−10
	Flowers	1	1·5

row minima is equal to the smaller of the column maxima, the game has a saddle-point, and the players should always adopt the pure strategy indicated by the intersection. In this game, the husband should always take home flowers!

The search for and identification of saddle-points is an important feature of game theory; and the chance that a matrix chosen at random will contain such a saddle-point is high for small matrices. Where, however, no saddle-point occurs, it can be shown that the game should be played by adopting a mixed strategy. This means that two or more of the strategies should be played, and that the probabilities with which each strategy is played can be calculated from the matrix. For each individual time the game is played, the choice of strategy should be at random but at fixed probabilities for the relevant strategies.

There is, of course, an obvious objection to regarding Nature as a malignant opponent seeking to minimize the outcome for a human, plant or animal opponent. Nevertheless, in situations in which we do not yet know enough about the response of organisms or our environment to select strategies which will yield the highest outcome at least on average, there may be some advantage in selecting a combination of strategies which is conservative in the sense that it minimizes the harm that can be done by the worst strategies that Nature can play.

The example quoted above is taken from Williams.[95] This book is an excellent introduction to game theory, written in an interesting and humorous way, with a wealth of practical examples. The simple ecological example below is a modified version of one given in this book.

Example 7.2 – Angling and feeding strategies.

A species of surface-feeding fish depends mainly on three species of winged invertebrates. These invertebrates are not equally common on the surface of the water. If the three species are characterized as x_1, x_2 and x_3, their relative frequencies are $15n$, $5n$ and n. In other

Table 7.2   Game matrix for angling and feeding strategies.

Strategies		Angler uses: x_1	x_2	x_3
	x_1	-2	0	0
Fish feeds on:	x_2	0	-6	0
	x_3	0	0	-30

Table 7.3   Game matrix for modified angling and feeding strategies.

Strategies		Angler uses: x_1	x_2	x_3	Lure
	x_1	-2	0	0	-1
Fish feeds on:	x_2	0	-6	0	-3
	x_3	0	0	-30	-15

words, there are 5 times as many x_2 as x_3, and three times as many x_1 as x_2.

If an angler uses one of these three species of invertebrates in catching the fish, by threading one on to a hook, the Table 7.2 shows the outcome, from the point of view of the fish, for each of the possible feeding strategies in relation to the bait being used.

It is possible to show that this game matrix has no saddle-point, and that both the angler and the fish should use the same mixed strategy of either using or feeding on x_1, x_2 and x_3 in the ratio of 15:5:1. The value of the game to the fish is negative and equals $-10/7$, indicating that, in the long run, the fish will be caught, but that the optimum feeding strategy will reduce the chance of being caught on any particular occasion.

If we now suppose that the angler switches to using a lure which may be mistaken for any of the three invertebrates, but which is twice as likely to arouse the suspicions of the fish, Table 7.3 represents the new game matrix. The new matrix still has no saddle-point, and the optimum mixed strategy for the fish is now in the ratio 3:1:0—the x_3 are now too dangerous to eat! For the angler, the optimum mixed strategy has the ratio 7:2:0:1 and he, too, should never use x_3 as a bait. The value of the game for this fish is $-30/20$, which is a slight reduction on the original game.

Game theory models represent an interesting and so-far little explored alternative approach to the solution of strategic problems. The extension to the more complex non-zero-sum games and to many-person games in which coalitions can be formed between the players represents a field of research which deserves increased attention, particularly in ecological research related to the assessment of environmental impact and environmental planning.

CATASTROPHE THEORY MODELS

The theory of catastrophes is an elegant development of mathematical topology applied to systems which have four basic properties, namely bimodality, discontinuity, hysteresis and divergence. Bimodality refers to the system being characterized by one of two (or more) states, while the property of discontinuity implies that there are only relatively few individuals or observations which fall between the two states. The characteristic division of organisms into males and females is a good example of both bimodality and discontinuity. The presence of occasional individuals whose sex is indeterminate does not greatly hinder the recognition of the two states, but the theory implies some discontinuity between the states, such that any individual will be readily placed in only one of the categories. Discontinuity also refers to any large change in behaviour or state associated with a small change in some other variable, including time. Hysteresis occurs when a system has an apparently delayed response to a changing stimulus, and characteristically the response to the stimulus follows one path when the stimulus increases and another path when it decreases.

Divergence is a more difficult property to describe, but is characterized by nearby starting conditions evolving to widely separated final states. In applications of these models to population dynamics, for example, initial conditions just above and just below quite well-defined thresholds frequently diverge to very different final states.

The simplest kind of catastrophe is illustrated by Fig. 7.2, which represents what has come to be known as the ***fold catastrophe***. Initially, the system is assumed to be at the point A on the lower branch of the fold manifold. As the variable p increases, the variable x also increases, so that the system passes through the point B until it reaches the point C. At this point, the variable p crosses the singularity S_1, and the system makes a 'catastrophic' jump to the upper branch of the manifold at C'. Further increases in the variable p carry the system to and beyond the point D.

If, however, the variable p begins to decrease, the system continues to follow the upper branch of the manifold through E to F. At F, the variable p crosses the singularity S_2, and the system makes a 'catastrophic' return to the lower branch of the manifold at F'. Thereafter,

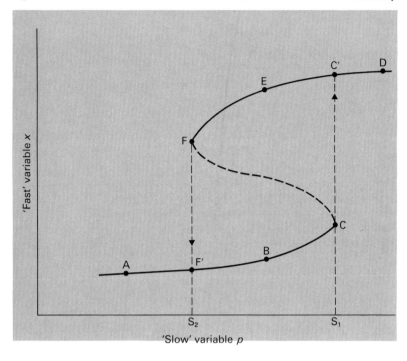

Fig. 7.2   Representation of fold catastrophe.

further changes in the variable p move the system towards either A or B until it again crosses the singularity S_1.

The simple fold catastrophe illustrates quite well the properties of bimodality, represented by the two branches of the fold manifold, and discontinuity, represented by the sudden transition from one branch to another at the singularities S_1 and S_2. Hysteresis is illustrated by the property of following a different return path after the crossing of a singularity. It should perhaps be noted that the specific form of the function relating x to p is not important as long as there is a fold singularity in the projection of x on p. The simplest polynomial function that is equivalent and representative of the fold catastrophe is

$$f(x:p) = -(x^3 - x + p)$$

The more complex *cusp catastrophe* is illustrated in Fig. 7.3. This system is assumed to be represented by a variable x which is dependent upon two variables p and q. Because of the fold in the surface representing the dependence of x on p and q, the behaviour of the system varies

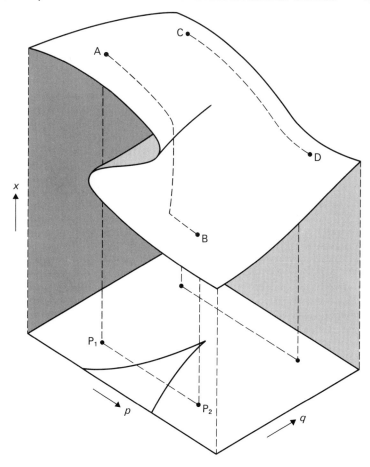

Fig. 7.3 Representation of cusp catastrophe.

according to the values of p and q. If, for example, p goes from P_1 to P_2, the system moves from A until it encounters the singularity and then makes a catastrophic jump to the lower surface before moving to B. If the system moves from C to D, however, a similar change in the value of p does not encounter the singularity. Whether or not the singularity is encountered depends on the relative values of both p and q.

Figure 7.4 illustrates divergence in a cusp catastrophe, where the system is shown in two nearby states E and F. If the value of q is

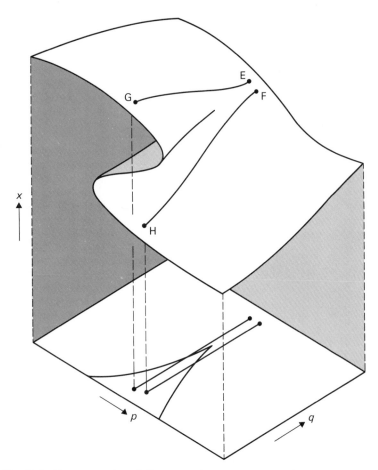

Fig. 7.4 Representation of divergence in a cusp catastrophe.

reduced, the system moves steadily to the points G and H respectively. Even though both paths start arbitrarily close, and both experience the same change in the parameter q, they end at widely separated final states. Because of the existence of the cusp, the paths of the two changes diverge, that of EG ending on the upper sheet of the manifold, and that of FH ending on the lower sheet of the manifold.

Anyone wishing to find a simple, but comprehensive, introduction to the application of catastrophe theory to ecological systems can hardly do better than to read the report by Jones.[40] In addition to providing a simple introduction to catastrophe theory, and defining the principal conditions required for its application, he illustrates the use of the theory in modelling the spruce budworm system of eastern Canada.

Example 7.3 – Dutch elm disease.

As an interesting, but highly speculative, example of the possible application of the catastrophe theory model, we may consider the outbreak of Dutch elm disease in Britain. This disease, which causes die-back and death of elm trees, is caused by the fungus *Ceratocystis ulmi* and is carried from tree to tree by bark beetles of the genus *Scolytus*. The disease was first recognized in Britain in 1927, although it was almost certainly present before this date. Since 1927 there have been several serious outbreaks of this disease during which it clearly changed from an 'endemic' to an 'epidemic' state.

Recent research has provided evidence that at least the present epidemic outbreak of the disease is due to an aggressive strain of *Ceratocystis ulmi* and it is postulated that the markedly regional distribution of the disease is the result of separate introductions of the aggressive strain in different places. Slightly affected trees, particularly in areas remote from the main disease centres, are thought to have the non-aggressive strain which forms the relict of the *Ceratocystis ulmi* population after the decline of previous epidemics.

An alternative hypothesis, however, is that the death of a large proportion of the infected elms and the presence of the aggressive strain are both reflections of the epidemic state of the disease, and that these are related to the population of Scolytid beetles and the density of elms per unit area, as in Fig. 7.5. A typical path for the development of an epidemic begins at A with a large number of elms per unit area and low population of beetles. If the beetle population increases, perhaps because of a run of mild winters, the path intersects the fold singularity at T, and then enters the epidemic state at B, exhibiting both the aggressive strain and the deaths of a large proportion of the infected trees.

The path from B has two distinct possibilities. If the number of elms per unit area is sufficiently decreased by the epidemic, or by selective felling, and the population of beetles also decreases, the disease returns smoothly to endemic levels. If, however, the population of beetles is reduced quickly, by treatment or by adverse conditions, before the number of elms per unit area is reduced sufficiently, the path of the disease is carried back over the cusp, giving a jump return to endemic levels at C, delayed by the extent of the fold.

Relatively little evidence has so far been published on the variation of populations of Scolytid beetles. Gibbs and Howell[27] have, however,

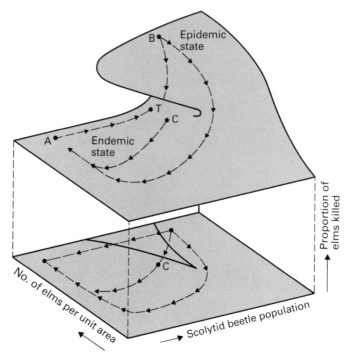

Fig. 7.5 Catastrophe theory model of Dutch elm disease.

given estimates of the numbers of elms per sq km in the southern counties of England, together with estimates of the proportions of infected elms and the proportions of severely diseased elms in 1971. If we regard the proportions of infected elms as a rough measure of the size of the beetle populations, a preliminary test of the catastrophe hypothesis can be attempted, and the relationships between the proportions of severely diseased elms and of infected elms for counties with less than 200 elms km^{-2} and more than 200 elms km^{-2} are plotted in Fig. 7.6. The data are not wholly inconsistent with a hypothesis of a manifold with a cusp at a point somewhere between 100 and 200 elms km^{-2} and 10 per cent of elms infected. Clearly, however, further data and analysis would be necessary to test the hypothesis.

Catastrophe theory models have attracted a lot of interest and attention since they were first proposed in 1970. The models have considerable intellectual and visual appeal, but are not easy to apply to highly multivariate situations. There are also serious difficulties to be overcome in estimating the parameters of manifolds from ecological

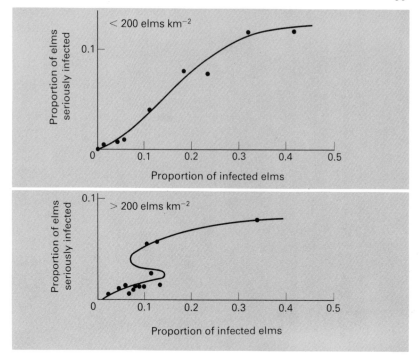

Fig. 7.6   Relationship between proportions of infected and seriously infected elms.

data. Nevertheless, we are almost certain to see further development and application of the models during the next few years.

This section concludes our review of the principal families of models available for the application of systems analysis to ecological problems. Any, or all, of the model families may be appropriate to a particular problem, and in the phase of generation of alternative solutions, as many of the models as possible should be explored at least in a preliminary way. In the next chapter, we will look in more detail at the modelling process itself.

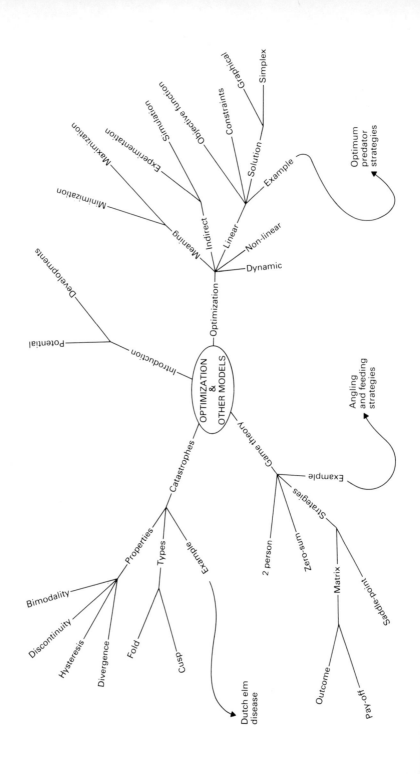

OPTIMIZATION & OTHER MODELS

Introduction
- Potential
- Developments

Optimization
- Meaning
  - Indirect
    - Maximization
    - Experimentation
    - Simulation
    - Minimization
  - Objective function
- Linear
  - Solution
    - Constraints
    - Graphical
    - Simplex
  - Example → Optimum predator strategies
- Non-linear
- Dynamic

Catastrophes
- Properties
  - Bimodality
  - Discontinuity
  - Hysteresis
  - Divergence
- Types
  - Fold
  - Cusp
- Example → Dutch elm disease

Game theory
- 2 person
- Zero-sum
- Strategies
  - Matrix
    - Outcome
    - Pay-off
  - Saddle-point
- Example → Angling and feeding strategies

8

The Modelling Process

In the earlier chapters, we have defined the basic philosophy of systems analysis and the use of mathematical models, and we have reviewed some of the families of mathematical models which are likely to be employed in the solution of ecological problems. Philosophy and examples of particular kinds of models are, however, still a long way from the practical application of mathematical models to real problems, and, in this chapter, we will consider the actual process by which the systems analyst has to understand the problem and work towards its solution. Mathematics may be an exact science, but, as we will see, the application of mathematics to real-world problems involves a process which calls for a high degree of intuition, practical experience, imagination, and what can only be called 'flair'. These qualities are especially necessary when, as is usually the case, the problem itself is relatively ill-defined.

DEFINITION AND BOUNDING

If we follow the phases of systems analysis identified in Chapter 1, our first task, once the existence of a problem has been recognized, is to define and bound the problem. Indeed, the primary focus of attention of our research at all times should be on this definition and bounding. Only in this way can we ensure that the limited research resources which we have to deploy will be correctly allocated and not dispersed into activities which are not relevant to the problem which we are meant to be addressing, and to the original universe of discourse.

Although there is a strong tradition for scientists to insist that their work should be unfettered by demands for relevance and for practicality, systems analysis does not belong to that tradition.

However, it is important to emphasize that our definition and bounding of the problem is unlikely to be correct at the first attempt (or even at the nth attempt!), and that there is, therefore, little point in aiming at perfection in one step. The main need is to make a start, preferably in the right direction, but if we subsequently find that we need to change directions, little will have been lost as long as we are prepared for the change, and as long as we have built up the necessary mental effort to sustain our initiative and momentum.

Our descriptions of ecological systems will usually need to be bounded by the constraints of space, time and sub-systems for which decisions have to be made. Statisticians have always stressed the necessity of defining the population about which inferences are to be made as a preliminary to any form of experimentation or sampling, and the argument is readily extended to the modelling of ecological systems. Our model is intended to facilitate inferences about some population, and our initial definition and bounding of the problem must be sufficiently explicit to identify that population.

An interesting, if localized, example of this difficulty is provided by research in the United Kingdom on the treatment of areas which have been covered by pulverized fuel ash, a waste product from coal-burning power stations. Pulverized fuel ash (PFA) is an almost inert material containing little or no organic matter and varying quantities of elements which are harmful to plant and animal life. As productive ecosystems have been covered with this material to a depth of several metres, our concern is to develop new systems which are capable of surviving and developing on PFA. Experiments on various forms of treatment and management will necessarily be concentrated on the earliest areas covered by PFA, but the relevance for later areas of the results derived from such research will be doubtful if, as seems likely, the industrial process or its raw materials change with time. The ecosystem in this case has been defined and bounded within space and sub-systems, but not in time. The systems analysis for the rehabilitation of such an ecosystem by the application of fertilizers and organic material, either as a surface dressing or incorporated in the surface layers of the material, and introduction of earthworms and other invertebrates will, therefore, require a sequential component to test the applicability of the results of earlier research to later material. The possible effects of climatic and weather cycles on rehabilitation treatments of damaged ecosystems impose similar constraints upon the design of investigations. Sequential techniques for such investigations and for such models have been available for many years, but are seldom used.

The bounding of the problem in space and time, however, is usually easier, and consequently more explicit, than the identification of the ecological sub-systems to be incorporated in the models. Many of the projects of the International Biological Programme (IBP) assumed that it was necessary to model the whole ecosystem and that it was therefore unnecessary to define the sub-systems of that ecosystem. When the final synthesis was attempted, it was found that in many IBP projects there were major gaps in the system which could not be filled by any of the experimental or survey results, and these gaps were frequently emphasized by the absence of any preliminary synthesis. The experience of IBP has led many ecologists to question the need for studies of whole ecosystems, and to focus attention upon carefully designed sets of sub-systems. In the synthesis of the tundra ecosystems, for example, most attention was concentrated on the decomposer and nutrient cycles as a basis for the prediction of the effects of environmental impacts upon the tundra.

COMPLEXITY AND MODELS

The application of systems analysis to ecology is relatively new, and few guides are therefore available for the construction of ecological models. As a result, untested hypotheses are frequently incorporated into model development, and the optimum number of sub-systems to be included in the model is difficult to predetermine for a defined and acceptable level of accuracy. It can be argued that a more complicated model should be able to account more accurately for complexities in the real system, but, while this argument may appear to be correct intuitively, there are some additional factors to be considered. For example, the hypothesis that greater complexity leads to greater accuracy has been tested by analysing the total uncertainty accompanying model predictions. In general, systematic bias resulting from abstracting the system into a few sub-systems is inversely related to complexity, but there is an associated increase in uncertainty due to errors of measurement of individual parameters within the model. As increasing numbers of parameters are added to the model, these parameters have to be quantified in field and laboratory experiments, and the estimates of the parameter values are never error-free. If these errors of measurements are carried through into a simulation, they contribute to the uncertainty of the predictions derived from the model. For all of these reasons, there is a very great advantage in reducing the number of sub-systems to be included in any one model.

Some ecologists, however, have emphasized the importance of niche structure in the dynamics of ecosystems, and suggest that an ecosystem model which ignores species differences runs the risk of neglecting important elements in its dynamics. Unless circumstances permit direct comparison of a simplified model with the observed behaviour of a representative range of ecosystems, acceptance of the simplified model should be based on a demonstration that deviations from the behaviour of an alternative model which takes biological diversity fully into account are negligible for the purpose in question. The trade-off between complexity and simplicity in the choice of systems and species to be included will be one of the most difficult problems that the systems analyst will face in any single practical application. As suggested above, he is unlikely to get it right in his first attempts at bounding and defining the model.

IMPACTS

Perhaps even more important than the need to define the levels of complexity of the sub-systems is the need to define impacts to be made upon the system. No model or research investigation can possibly foresee all, or even most, of the ecological impacts, and any investigation will need to be qualified by a series of hypotheses about the relevance of managerial treatments and impacts. Ideally, the basic structure of the investigation will enable the interaction of the various factors to be investigated. Where experiments are to be conducted, the design of the experiments can incorporate the factorial structure of the impacts in such a way that the effects of the impacts are not themselves confounded. The many devices for the control of factorial structures in experiments enable such investigations to be carried through with economy and precision. Even where direct experimentation is not possible, however, it is still necessary to enumerate the relevant impacts and to sample the defined system so that the effects of these impacts are, if at all possible, unconfounded. There may, indeed, be little point in continuing an investigation for which it is not possible to separate the effects of two or more impacts.

A particularly pertinent example of this difficulty can be observed in current attempts to model the relationships between the variability and concentration of acid rainfall on tree growth. With some difficulty, it is possible to measure short-term fluctuations in the growth of individual trees, but, so far, only a relatively small proportion of the total variability of growth in a period of, say, one hour can be accounted for by climatic variables, including temperature, moisture, wind speed and evapotranspiration. When the possible lags between the variations in climate and the growth response of trees are considered, however, it is not surprising that simple growth models are not markedly successful. If, however, we add the fluctuations of concentrations of acid rainfall to this model, we face the further difficulty that the acid rainfall is itself closely correlated with the same climatic variables used to characterize the growth response of the trees. It follows that recording of growth and climate at a limited number of sites is unlikely to provide any useful information on the complex interaction between climate, acid rainfall and tree growth, unless some way can be found of separating the confounded effects of climate and acid rainfall.

We have, perhaps, dwelt so long on the difficulties of making a start on the bounding and defining of the problem that most investigators will now have been deterred from even starting! However, it is preferable to face the complexity of the problem in its full extent rather than to attempt to solve some minor part of the problem and then assume that we have solved the whole. One of the besetting sins of systems analysts is for them to abstract a part of the problem convenient for solution, and then to pretend that the part which they have chosen is the whole system.

WORD MODELS

Having made the first attempts at bounding and defining the problem, most analysts will then probably seek to establish a word model of the problem that they are seeking to solve. We have already discussed the desirability of such models, and indeed the validity of calling a verbal description a model of any kind. Nevertheless, before embarking on some mathematical solution, it will seldom do any harm if the mathematician or systems analyst writes down what he understands the problem to be in as simple words as possible, and agrees this description with the ecologists and with the managers who, he hopes, will ultimately use his systems analysis and models. Much subsequent discussion can be avoided if this simple precaution is taken. Of course, the systems analyst will doubt whether mere words will be able to capture the subtlety of the final model which he hopes to produce, and the ecologist will often wonder whether the mathematics is going to define the problem any better than the words—it is, however, in the exploration of this interaction that the richness of the scientific investigation will be expressed. Much of the difficulty will reside in the fact that the various groups of people concerned with the solution of the problem will seldom be able to communicate their ideas at all freely, whether they use words, mathematics, diagrams, or just plain intuition. It is best, therefore, to use as many of these modes of expression as possible.

GENERATION OF SOLUTIONS

Having got this far, we are now in a position to make a list of possible alternative solutions. How many solutions should we seek? There is no simple answer to that question, except, perhaps, to suggest that the would-be analyst should first write down as many different approaches to the same problem as he can imagine. On further reflection, the difficulties, mathematical or conceptual, will make some solutions seem unpromising, but the alternatives should not be neglected for these reasons, at least at first. It may also be found subsequently that many of the alternative solutions proposed require information which is not available, or which cannot readily be collected. Again, alternatives should not be rejected too quickly for this reason. It is often possible to discover that data which were not known to exist are actually available from some unexpected source, and the necessary variables can often be derived as secondary computations from other data, at least as a first approximation. Even the most unlikely solutions should therefore be listed in the first review of possible alternatives. Frequently, it will only be after thinking about the possible solutions over several weeks that the opportunity to exploit some mathematical convenience will become apparent—the human mind works slowly and with unpredictable barriers and lags!

Once a list of the possible solutions has been made, the next stage is to consider the ways in which these solutions might be combined in two, three or more ways at a time. Such combinations may appear unlikely in the early stages of the systems analysis, but the opportunity to link diverse methods of approach should not be lost, and, again, will become apparent only after the mind has been allowed to dwell upon the possibility for some considerable period of time. The actual linking may, indeed, become possible only after work has started on several of the models. The important need is to prepare the mind for a wide range of alternative solutions and for their combinations.

HYPOTHESES

In essence, any investigation of ecosystems which purports to be scientific, and which attempts to use systems analysis as the method of investigation, requires the definition and bounding of the problem to be framed as hypotheses which can be tested formally, even if that test can only be conducted after a chain of deductive reasoning from one or more hypotheses that are incapable of direct verification. Three basic classes of hypotheses may be distinguished, and will be used in the implementation of the various alternative solutions that we have listed above.

1. Hypotheses of relevance, identifying and defining the variables, species and sub-systems which are relevant to the problem.
2. Hypotheses of processes, linking the sub-systems within the problem and defining the impacts imposed upon the system.
3. Hypotheses of relationships, and of the formal representations of those relationships by linear, non-linear and interactive mathematical expressions.

These three classes of hypotheses may well be linked within a formal chain of argument, leading to processes which can be summarized by a decision table enumerating all the hypotheses, and combinations of hypotheses, that must be specified in order to solve a particular problem. The decision table also identifies, for each combination, the decisions or actions that need to be taken to ensure that the problem is correctly solved. Because decision tables provide a clear and concise format for specifying a complex set of hypotheses and the various consequent courses of action, they are frequently ideal for describing the conditions for interaction between component parts of the model. The extension of these techniques to the enumeration of the necessary combinations of hypotheses for particular courses of action where uncontrolled events may intervene, so that we are unable to predict or control with certainty, has been the main thrust of recent research into decision analysis.

This emphasis upon the necessity for formal hypotheses defining and bounding the problem for investigation should be sufficient to dispel any lingering impressions that systems analysis is some form of higher magic through which the abstractions and algorithms of mathematics will enable problems to be solved without careful thought! Indeed, the need to formulate hypotheses so that they are capable of being tested—usually by statistical techniques—will itself focus the major research effort on to logical thought rather than on computation, mathematics and computers. If that thought has insufficient logic, no amount of computation will rescue the model from inevitable failure, no matter how enjoyable the computational exercise.

MODEL CONSTRUCTION

We now come to the stage of the analysis which systems analysts most enjoy, i.e. the actual construction of the model and the mathematical manipulation of the various ideas which have been expressed in the hypotheses. Data will be gathered and examined carefully to test departures from the formulated hypotheses. Graphs will be plotted, either manually or on computers, to examine relationships and to determine whether these relationships are linear, non-linear or inter-active. Existing sets of data will be scanned for outlying individuals, and varying statistical tests will be tried, many tests sometimes being carried out on the same set of data. To the non-mathematician, and indeed to the mathematician himself, much of this activity will look like 'play'. It should be obvious that most applied mathematicians will be thoroughly enjoying themselves during this phase, and will be reluctant to draw it to a close, or, sometimes, even to review how far they have got in trying various alternative solutions to the original problem. They may even forget what the original problem was unless they are firmly reminded from time to time by reference to the word model, and to the definition and bounding of the problem. There will be endless temptations to explore all kinds of exciting ideas which are peripheral to the main purpose of the investigation. There will be all kinds of reasons why any given time will be inopportune to sum-marize the results of all this activity in some coherent form and to evaluate the outcome. To the mathematician it is fun, and it is right that it should be fun!

Nevertheless, there must come a time at which a choice is made of the number of alternative solutions which are to be carried through to the final solution of the problem. For all kinds of reasons, some of the alternatives will have dropped out of consideration. Perhaps the mathematics became too difficult; perhaps the data were not available; perhaps what seemed like a good idea when it was first examined is no longer adequate when looked at in relation to the available information, and to the purposes of the original problem. Inevitably, two or three, or perhaps even only one, of the original alternatives and their com-binations now seem a feasible approach to the solution of the original problem. It is then time to draw together the resources of the investi-gation, check the definition and bounding of the problem again, as it will almost certainly have been modified during the active and hectic phase of actual mathematical work, and draw together the conclusions which have been derived from all this computation, graph plotting, and plain hard thought. We now have to approach what is perhaps the most difficult phase of the whole exercise which is the verification and validation of the systems model.

VERIFICATION AND VALIDATION

Our emphasis on hypothesis formulation will help to clarify the distinction between verification and validation. Although the usage of these terms is not consistent, verification may be regarded as the process of testing whether the general behaviour of a model is a 'reasonable' representation of that part of the real life system which is being investigated, and whether the mechanisms incorporated in the model coincide with the known mechanisms of the system. Verification is, therefore, a largely subjective assessment of the success of the modelling, rather than an explicit test of the hypothesis underlying the model. To a large extent, therefore, some verification will inevitably have been going on during the hectic phase of mathematical activity, as the 'reasonableness' of the results will be one of the criteria by which the modeller will have judged the success or failure of his efforts. Nevertheless, what is reasonable in small parts of the model may be less reasonable when those parts are put together into a composite of the individual parts. Interactions between responses and impacts may need to be explored sequentially and factorially to ensure that the full range of possible conditions has been covered and that, within the limits bounding and defining the problem and the ecosystem, the model behaves, for the defined purposes, in much the same way that the real system behaves. We must, of course, be careful that we do not reject a model simply because it behaves in a counter-intuitive fashion. There are plenty of examples of solutions which are contrary to what is usually regarded as commonsense. No model should be rejected, therefore, simply because the results are unexpected. Where, however, the model behaves in a completely different fashion from the real system which is being investigated, some explanation has to be sought, at the very least, for the inconsistency. This is the role of verification.

Validation, in contrast, is the quantitative expression of the extent to which the output of the model agrees with the behaviour of the real-life system, and is the explicit and objective test of the basic hypothesis made by means of a delineation of test procedures, primarily statistical, which are applicable to the determination of the adequacy of the model. In most ecological applications of systems analysis, this process of validation has hardly been attempted, mainly because of inadequate definition and bounding of the problem initially. Typically, validation, where it is attempted at all, is approached in a direct and obvious manner, mainly by observing the behaviour of the model systems under a set of controlled or measurable loading and other conditions and then comparing the observations to corresponding predictions of the simulator. When the observations and predictions agree within required limits for all conditions treated, the simulator is considered to be validated.

This procedure has several recognized difficulties, not the least of which is the uncertainty associated with drawing a general conclusion from a finite (and typically small) number of experiments. This uncertainty is of particular concern in the validation of systems analysis models, where one may be attempting to predict effects which are of the same order of magnitude as the random fluctuations or 'noise' inherent in real system measurements. In such cases, it is advantageous to use techniques for statistical design and analysis of experiments, both to reduce the number of experiments needed for a given level of confidence and to indicate the statistical significance of measured and simulated effects. Fortunately, effective techniques for experimental design have been developed during the last fifty years, and techniques which were originally intended for use in experiments on real-life systems are now proving valuable in testing the behaviour of simulators of those systems. A full account of these techniques is outside the scope of this book, but the interested reader will find useful reviews in the papers by Schatzoff and Tillman[68] and Kleijnen.[43]

The most important aspect to be included in the validation of the models is the deliberate variation of more than one parameter at a time. Unfortunately, many research scientists have been educated, quite erroneously, to the belief that all good scientific work operates by the changing of only one factor at a time. As R. A. Fisher pointed out in the 1920's, such research can never adequately investigate the interactions of two or more factors. It is almost certain, therefore, that the validation of systems models will require factorial experiments to determine the effects of varying levels in the model parameters, and the interactions of the changes in these parameters. Where the number of parameters is large, fairly complex experimental designs which involve only some of the large number of factorial combinations of these parameter changes will be useful. In contrast to field experiments, experiments on the models in systems analysis are carried out sequentially, so that many forms of sequential sampling and sequential experimentation, possibly involving cyclic designs and response surface evaluation, are particularly appropriate. In other cases, techniques appropriate to experimentation on pilot plants in industry, such as evolutionary operation, are also appropriate to the validation of the models of systems analysis. Good accounts of all these techniques are available in standard texts on the design of experiments, e.g. by Davies[15] and Cochran and Cox.[12]

SENSITIVITY ANALYSIS

The investigation of the effect of changes in the input variables and parameters, and whether these changes produce large or small variations in the performance of the model, is sometimes known as sensitivity analysis. Ideally, such analysis should begin as soon as any modelling is attempted, and form part of what has been referred to above as the 'hectic phase' of modelling. Parameters to which the model behaviour is sensitive can then be made the subject of close scrutiny and subsequent modification, and it may then be necessary to undertake further experimental work or data analysis to ensure that those mechanisms are more precisely modelled. Sensitivity analysis, particularly if carried out early in the research project, may greatly aid decisions about the allocation of resources to various parts of the research programme.

Uncertainties in model performance can also be investigated by sensitivity analysis and, because actual uncertainties in the knowledge of each parameter can be estimated, experimental variation in the appropriate order of magnitude may deliberately be introduced into each run. For large and complex models, sensitivity analysis and the testing of the validity of the model as a whole can be a long and expensive process, but it is essential to discover how models behave within the full range of variation of the basic parameters, and studying the effects of change in one parameter at a time provides no information about interactions.

PLANNING AND INTEGRATION

As a consequence of the development of the techniques of systems analysis, and the use of computers and computer languages (a topic which will be dealt with in the next chapter), it is now possible to design research strategies for ecological investigations having many interacting components. Interdisciplinary groups are increasingly bridging the gap between disciplines and methodologies in many research programmes. These programmes are often directed towards the solution of problems which are likely to be of critical importance in the long-term, but which are of limited practical importance in the short-term. On the other hand, management agencies are continually embarrassed by the need for short-term, limited solutions to immediate problems, and have therefore evolved a strong bias towards pragmatism. Synthesis and integration of research and management attitudes are needed to combine the precision and detachment of the research with the pragmatism and immediacy of the management agency, in order to help develop new approaches to problem-solving. Experience of systems analysis in ecology and in other fields suggests that the formulation of hypotheses and collection of data, and the planning and implementation of pilot studies and management plans have very often been undertaken by different groups and by different agencies. Inadequate integration of these tasks has resulted in a loss of communication:

(a) between data experimentation and model development;
(b) between simulation models, the overall systems analysis and the implementation of models in management practice;
(c) between the examination of predictions from systems analysis and the implementation of models in management practice;
(d) between management practice and the development of new hypotheses;
(e) between the implementations of results from pilot studies and the development of new hypotheses.

The development of models often follows a somewhat standardized pattern, in which there is a progressively detailed breakdown of each component into modules which can be more easily translated into research activities. The following conclusions can usually be drawn.

(a) The quality of the available data and the understanding of causal pathways, especially as they relate to ecology, are generally poor.
(b) Systems analysis and data collection must develop a mutual feedback from which the decision-maker can draw maximum benefit.
(c) Training in systems analysis is valuable for stressing a broad, interdisciplinary, problem-oriented philosophy of research.
(d) Systems models can be improved only by building them and striving to correct their weaknesses.
(e) Systems analysis teams must be broadly interdisciplinary.
(f) Research which does not use systems analysis may demand large quantities of high-quality data and may consequently be expensive.

This chapter can do little more than indicate the main steps in the whole process of modelling. Nevertheless, every systems analyst and modeller probably operates in a slightly different way, and places different emphasis on the various phases of the development. I have tried to emphasize that the whole process is, or should be, controlled by the initial definition and bounding of the problem, and that systems analysis is not a group of techniques or methods in search of practical applications. The methods exist, some mathematicians have a wide experience in their application, and, obviously, differing preferences and expertise, and it is the purpose of the systems analyst to exploit such information as we have about the ways in which mathematical models behave in the solution of ecological problems. As we have seen earlier in this book, one may begin by trying to define the ecological system in as much of its complexity as can be encompassed by the human mind, and then find mathematical expressions for this complexity. The disadvantage of this approach is that one quickly finds that the model itself is too complex for rational thought and manipulation. The alternative is to accept some reduction in the level of complexity which will be permitted in the model and to exploit the mathematical convenience of well-developed logical systems. The process by which this modelling is done may sometimes appear chaotic to the outside observer, and especially to the non-mathematician, but this process does follow clearly recognized phases which can be used to direct the very considerable scientific and intellectual activity towards a successful and pragmatic conclusion. In the next chapter, we shall examine the role of the computer and computer languages in the modelling process and in systems analysis generally.

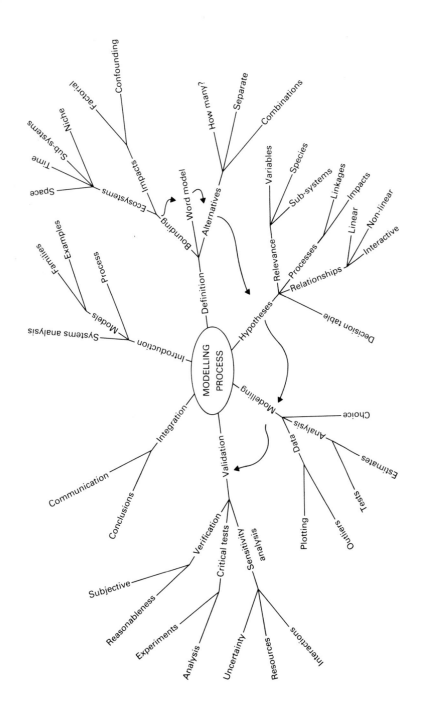

9

The Role of the Computer

Although little specific mention has so far been made of computers in systems analysis, it will, I think, be clear that the use of computers is implicit in the description of systems analysis and model families in earlier chapters, and in the account of the modelling process in the last chapter. However, attention has so far been deliberately focused away from computers to avoid any suggestion that it is the use of computers which distinguishes systems analysis from other forms of practical investigation and research. There are, nevertheless, three basic reasons for using computers in systems analysis, namely:

(a) their speed of computation;
(b) their ability to provide rapid access to large quantities of data;
(c) the value of algorithmic languages in communicating information.

Let us consider each of these three reasons in greater detail.

The fact that computers perform calculations at almost incredible speeds is, of course, well-known, and will be the most obvious reason for using a computer. It is not easy to convey this speed of computation to anyone who has not had practical experience of computers—something which adds fuel to the unceasing debate on computers in everyday life. However, if we think of the amount of computation a normal person could do in a year with just a pencil and paper and a knowledge of the usual multiplication tables, assuming no time for food, sleep or relaxation, the same amount of work can be done in two or three months with the aid of a simple calculating machine, such as can be purchased in any good stationery shop. Even the slowest of the electronic computers now available will do the same amount of work in minutes, and the most powerful of today's computers will do that amount of work in seconds. Without a computer, almost all of the necessary computation in systems analysis would be impossible.

Why do we need these almost incredible speeds of computation? In the first place, we need them quite simply because many of the mathematical techniques we want to use become impossible without the ability to compute quickly and accurately. Computations like matrix inversion, matrix multiplication, and of eigenvalues and eigenvectors are both tedious and time-consuming if attempted by hand or by the use of simple calculators, and the possibilities of making errors which are not readily detected until the computations are nearly complete, if then, are manifold. The fitting of even quite simple curves or regression relationships quickly becomes impossible to contemplate when one realizes that even one such calculation may take several weeks. Without the computer, therefore, many of the mathematical methods we want to use would not be practically feasible, even though we might be well aware of their theoretical applicability.

This access to a more powerful battery of methods is not, however, by any means the only reason for the importance of the computational speed of the computer. Although modern computers make few errors in the actual arithmetic of the calculations—and when such errors occur the computer itself tells us that they have occurred—the users of computers frequently make errors in the values assigned to parameters of models, in the assumed shape of the relationship between two or more variables, and in the amounts or quality of data subjected to analysis. Often, we can only recognize such errors towards the end of a computation, or when we examine the final results. If that computation has taken us several weeks to do, we are unlikely to view with any relish the prospect of having to repeat it with different parameters, assumptions or data. Knowing that we can repeat the whole operation quickly and painlessly in a matter of seconds encourages a closer scrutiny and checking of results at every stage of the investigation. It is only those who have never used computers for scientific research who believe that the data are put into the computer at one end, some calculations are done once by the computer, and the acceptable results come out at the other end!

One of the activities that is encouraged, or should be encouraged, by the speed of computation achievable with computers is the examination of our data, raw or processed, by the plotting of graphs or other techniques. With only small quantities of data, it is relatively easy for the research scientist to examine these data critically, plot graphs of one variable against another, look for outlying values which suggest the presence of anomalies or errors, and generally obtain the 'feel' of the data. Statisticians who were analysing data on desk calculating machines before computers became generally available—including the author—were aware of the value of having to key data into calculating machines, often several times, in detecting unusual values and obtaining an overall impression of the data. Handling large quantities of data quickly becomes impossible for the research scientist, or at least unproductive, on these machines, so that the input of data to computers and other forms of storage is delegated to assistants who may have little or no understanding of the purpose or theory of the analysis, and are therefore unable to obtain any critical view of the data. Fortunately, the speed of computation available through the use of computers enables the research scientist to examine his data in considerable detail before attempting any kind of formal analysis.

All of these considerations contribute to the fact, not appreciated by many people not engaged in scientific research, that nine out of ten research computations are 'wasted' in the sense that the computations are not actually used in the final results. They are, of course, not wasted in the sense that they have contributed to a greater understanding of the problem and its solution. This important distinction between research computing and automatic data processing, as in the calculation of a payroll or the control of merchandise in a warehouse, often has to be explained to computer managers for whom 'productivity' is measured by the proportion of 'successful' runs!

Speed of computation itself, however, though valuable, is less important than the second of the reasons for using computers in systems analysis—their ability to provide rapid access to large quantities of data. Use of a computer at all implies making our basic data machine-readable, that is the data have to be transferred to some medium which is capable of being read by the computer. The simplest, and earliest, of these formats is the punched-card, containing 80 columns upon which a variety of numerical and alphabetical characters can be punched. These cards can then be read by a computer and transferred to some convenient form of storage within the machine. An alternative format is punched paper-tape on which numerical and alphabetical characters can be punched, and, again, the paper-tape read by the computer. There are various advantages and disadvantages of punched cards and paper-tape, although the choice between the two media is more usually made by prejudice than by logic. Increasingly, and especially as the interactive use of computers has become more readily available, data may be keyed directly on to one of the main forms of internal computer storage, i.e. on to magnetic tape or magnetic discs.

Whatever form of input is used, once data have been read by a computer, or put into a form capable of being read by a computer, and checked carefully, they are then available for any appropriate analysis. They can be exchanged with other research workers, transferred to another computer, either directly, or through one of the available media of punched-cards, punched paper-tape, magnetic tape, or exchangeable magnetic discs. The data can be edited, put into several alternative formats, made available but protected from any change, or even protected so that they cannot be used or altered without prior permission. Sorting and other clerical operations can be carried out on the data without any need to enter them again, so that, once the data have been checked, errors are unlikely to creep into them. The worst that can happen is that the data may be lost through some malfunctioning of the computer, but it is usual to arrange for data to be held in at least two separate forms and locations as an insurance against this eventuality.

The ease with which data can be held and accumulated within a computer has led to the development of the concept of the 'data bank'—interpreted as a bank of information available to a wide range of people, the subscribers to or customers of the bank. This concept needs to be treated with some care! In the first place, all data are held within a computer in a pre-determined format. It is, therefore, essential that everyone wishing to use data in the 'data-bank' should be aware of the format and its constraints. For example, data on the numbers of organisms which have a particular disease may have been collected for only a small sample of the total number of organisms, and the 'percentage of diseased organisms' therefore may have a low precision. If these percentages are subsequently compared with data with a very different level of precision, the comparison may be misleading. Even more important is the possible effect of constraints on the way in which the data were originally collected when these constraints are not held with the data themselves.

We also have to be careful not to fall into the trap of believing that, just because the data are there, we must use them all. It will often be the case that we need only a fairly small sample of the data available in the computer for a particular purpose, and we should use the very extensive theory of sampling which has been developed particularly during the last fifty years. A small, properly-chosen sample will be easier to handle in our computations, will generally reduce computing times still further, and will often greatly clarify both the argument and results of the investigation.

There is another good reason for working with samples of data wherever possible. Because it is logically impossible to formulate and test hypotheses on the same set of data, it will usually be preferable to analyse in detail only a sample of the whole data, and to derive new hypotheses from this analysis. These new hypotheses can then be tested objectively on the remaining data. If the data set is really extensive, this process of analysis, hypothesis formulation and retesting of hypotheses can be extended sequentially to provide a powerful and logically correct solution to ecological problems.

The third reason for using computers in systems analysis is, in fact, by far the most cogent. Because computers can only be used if they are directed by a precise and completely unambiguous set of instructions, those instructions provide an accurate and logical account of the computation and clerical operations which we have used in analysing our data and constructing our models. The instructions are called, in computer jargon, a 'program'—the use of the North American spelling is quite deliberate, and, in Europe anyway, helps to distinguish between this special use of an ordinary word and the more general 'programme' of events. To most beginners in the use of a computer, the need to write a program, or programming, is a nuisance, but one quickly realizes the extraordinary advantage of having an unambiguous description of the computations. If programs are properly annotated when they are first used, it is often possible to discover exactly what was done during a computation months, or even years, earlier.

In the early days of the development of computers, programs were necessarily written in what were called 'machine-languages', and these languages were specific to a particular type of machine, and sometimes to an individual machine. Programs in machine-code were time-consuming and difficult to write, and were not readily transferred from machine to machine, and therefore from person to person, unless they happened to be working on the same computer. Fairly quickly, however, it became possible to write programs in simpler ways, at first in what were called 'autocodes' and then in several high-level languages. The best-known of these languages are undoubtedly FORTRAN (standing for FORmula TRANslation) and ALGOL (standing for ALGOrithmic Language), and a very large number of all computer programs for scientific computations are written in either of these two languages. More recently, the language of BASIC (Beginner's All-purpose Symbolic Instructional Code) has been added to the list of commonly used languages, its main advantages being that it is easy to learn and that it is particularly appropriate for interactive use of computers through terminals.

Some computer users advocate the use of special-purpose languages for modelling, including DYNAMO and CSMP, both of which simplify the programming of dynamic models. While it is certainly true that some limited families of models can be programmed for computers by means of such languages, in general the user would need to know a different special-purpose language for each family of models, and, even if he knew these languages, might still find himself constrained by the facilities provided by the language.

Anyone who wants to work in systems analysis as described in this text would be best advised to do his computing in one of the high-level languages of FORTRAN, ALGOL, and BASIC. The languages are flexible, extraordinarily powerful, and provide access to a wide range of existing programs, sub-routines, and algorithms. Learning any of these languages (and preferably all three!) opens up a new and virtually inexhaustible storehouse of knowledge and techniques which can hardly be described. Furthermore, the experience of learning to use a computer itself transforms much that was previously tedious and pedestrian to a new world of insight and possibilities. Some have described the experience as the nearest thing to being struck by light on the road to Damascus!

Computer programs themselves constitute an important medium of exchange of information. Indeed, communication with another systems analyst is often best made by examining the way the programs for his models are written. If we want to know how he simulated biological processes like photosynthesis or respiration, or how he introduced genetic variability in the reproduction of an organism, we will find a complete and accurate description in his computer program. We may then use the same piece of program or modify it in ways which seem more appropriate to us. If we communicate the results of our experience to the original programmer, our models will be enriched by the sequential improvements of our approximations to the real world. The language of words is often inadequate to describe this experience, but the languages of computers can be implemented on the machines which also provide access to the data we need, and which are fast enough to perform almost any form of computation that we may decide is appropriate.

Computer programs also embody a considerable degree of flexibility in their use, particularly when they are well-written. The same program may be used many times with different data, and even for purposes for which it was not originally designed. The smaller, self-contained parts of a large program—called sub-routines by computer users—are often capable of being extracted from the original program and used in ways not envisaged by their authors, or combined in different sequences to perform new kinds of computations. This flexibility and the degree of control a programming language gives to the research scientist are difficult to describe to anyone who has not experienced them. Certainly, there are very few people who have experienced the revelation of learning to program a computer who would relinquish the power and knowledge that they have gained.

One final feature that well-written computer programs have is the ability to guide subsequent users of the program through the constraints that inevitably surround any computation or modelling in ecology. By introducing various tests of the data used in the computations, care can be taken to see that only data which are consistent with the assumptions underlying the model are used. Not all computer programs achieve this level of sophistication, perhaps, but it is perfectly possible for a skilled modeller to ensure that his model is not subjected to the two distortions most likely to wreck his reputation, i.e. extrapolation to situations for which the model was not intended, and the incorporation of wholly inappropriate kinds of data.

The main conclusion of this chapter is that the computer has an essential role in systems analysis and modelling. Indeed little serious work, especially in ecology, can be attempted without use of computers and anyone who wants to do research in this field will need to learn how to use one. Use of a computer without learning how to program one is like owning a Rolls Royce or an E-type Jaguar and having to depend upon a chauffeur! It is, therefore, essential to learn one or more of the high-level programming languages of FORTRAN, ALGOL, and BASIC. Learning the first of these languages will take about as long as it takes to learn to drive a car, if the language is to be used well, but the experience is worthwhile both from the point of view of being able to do work which could not otherwise be attempted and from the point of view of the access that is gained to the experience held in computer programs written by other people. The second computer language takes much less time to learn—certainly less than half the time needed to learn the first—and the third language progressively less.

The practical experience of programming will greatly increase the flexibility of the modelling which can be undertaken in any application of systems analysis to ecological problems. Models of many kinds can be developed, tried and, if necessary, abandoned without the systems analyst feeling that he has devoted so much time to the model that he is reluctant to see it scrapped. Indeed, one of the perennial difficulties in modelling and systems analysis lies in persuading research workers to abandon models which have outgrown their usefulness, particularly when they have spent months, or even years, developing them. The use of a computer for all of the computational effort required in analysis and modelling simplifies the approach to the task, and places the emphasis where it should be, on the conceptual aspects of the work.

In addition, the use of the computer greatly facilitates the exchange of information between research workers, and provides access to the massive volume of previous work. In the future, it seems likely that the results of research will increasingly be transferred through the models we build in systems analysis, and these models can be communicated most easily, and most accurately, through the computer programs that enable them to be implemented on computers. Instead of talking about 'data banks' we will talk about 'model banks', and our first approach to many new problems will be to see if we have any close analogues to those problems in our 'model bank', at least for a first approximation. Ideally, of course, we would like to build a new model for every problem, but the world's problems are so many and so diverse that there is little hope that we can work on more than a relatively small number of them from the start. The communication of what has been achieved in the past then becomes of the greatest importance.

One of the most rapid developments of science and technology is currently taking place in computers and computing. The computers themselves are becoming smaller, more powerful and cheaper to use. The kind of computer which is now called a 'minicomputer' is many times more powerful in operation than the machines which were available as large computers only ten years ago. They are also available at only a fraction of the cost of the machines which they have replaced. Today, even these minicomputers are being replaced by 'micro-computers' with the same capacity and speed. At the same time, really large computers are being constructed and these machines will enable tasks to be undertaken that could only be dreamed about even a few years ago. The combined use of the new small computers for our everyday computing with the really large computers when we need access to major data sets or extra rapid calculations will place even greater power in the hands of the system analyst and enable him to explore models which are more complex and more able to encompass the variability and the interactions of the physical and ecological world.

The rapid development of computers has also stimulated an equally rapid, and even more far-reaching, development in the algorithms necessary for the practical use of these computers. As man's intellect begins to encompass the possibilities revealed by the ability to compute and perform logical operations at speeds which, even a few years ago, were unbelievable, new ways of thinking about our world and our impact on that world become encapsulated in the algorithms we prepare for testing these ideas on computers. The history of programming languages, and of the applications of these languages to practical problems, is short, but it already contains some of the most important conceptual advances that man has yet made. In turn, these conceptual advances have led to a development of mathematics itself—not the classical mathematics which we teach in schools and universities, and certainly not the classical applied mathematics which is essentially mathematics applied to the physics of the eighteenth and nineteenth centuries. This new mathematics, made possible by the computer, but stimulated by the practical problems of our modern world, is, perhaps for the first time, capable of dealing with the dynamics, the variability, the uncertainty and the catastrophes of our world and of our universe.

However, this chapter should not finish without a return to the recognition that the systems analysis described in this book is a function of the human intellect. Although computers and computer languages make possible many of the steps in systems analysis, they are only tools to be employed by the intellect. They can be misemployed and misused as can any tool. No amount of computation can make up for faulty assumptions and incorrect logic. Accuracy of description of a model is worthless if the model itself is based on a false premise and a false logic, but the accuracy of the description may help another scientist to recognize the faults in the premises and the logic. Perhaps the greatest danger, however, is that the model, enshrined in the new 'magic' of the computer and its language, is too often taken for the real world for which it is an approximation. The models of systems analysis are, or should be, ephemeral, to be replaced by new models, new approximations, as our knowledge increases.

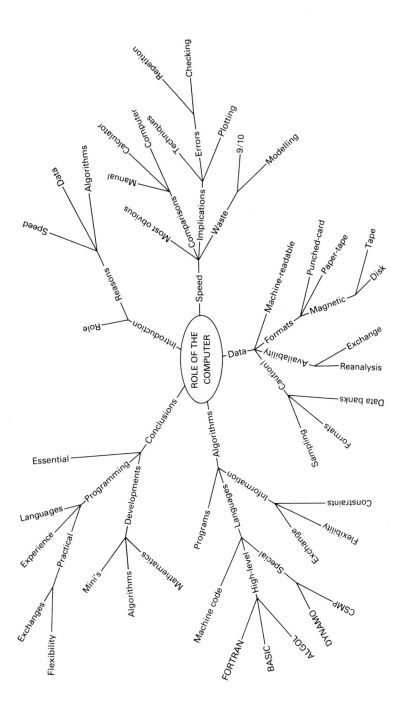

Postscript

This introduction of systems analysis in ecology has a limited purpose. It is intended to introduce the reader to the concepts of systems analysis and of mathematical models, and to explain why these concepts are needed in ecology. A large part of the book has been taken up with describing some of the more important families of mathematical models, with simple examples of applications to ecological research. We have ended with a review of the modelling process and of the role of computers in systems analysis.

The perceptive reader will, by now, be asking the question: 'But why haven't you taken one problem in ecology and shown the development of all the alternative models for that problem, how these models were selected and rejected, and how the implementation of the results of the analysis was proposed and followed up?' This approach was considered when the foundations of this book were laid, but, after much thought, was eventually rejected. In the first place, it would have made the book too long for an 'Introduction'. Second, the choice of problem would almost certainly have reduced the range of interests of likely readers, and much of the argument and discussion would have been transferred to the application rather than to the principles which this book has attempted to stress. Third, applications of systems analysis to practical problems are already beginning to be published, and some of these applications can readily be extended to show the way in which the general philosophy was developed and expanded to meet the special requirements of the problem.

For any reader interested in following up some of the aspects of systems analysis touched upon in this introduction, some further help may be desirable. First, it goes without saying that no ecologist can learn too much mathematics, and no mathematician can learn too much ecology, if he wants to work in the field of systems analysis. If you are an ecologist, start to read some texts on mathematics. There are hundreds to choose from, so treat mathematical texts as you would novels—if the book doesn't appeal to you, put it down, and look for one that does. Start with a relatively elementary text on algebra, matrix algebra, co-ordinate geometry, calculus or statistics, and then build up a solid foundation of knowledge. Don't make a chore of it— skip any bits that are too hard on first reading and go back to them later—read for the enjoyment of mastering new concepts.

If you are a mathematician, start to read some of the many hundreds of books on general or specialist ecology. Again, don't persevere with any book that bores you—you may come to appreciate it later, or you may decide it is not worth returning to it. You will be irritated by many ecologists' inability to use symbolic representations with the result that many of their books are much longer than they need to be. If you can see ways in which your mathematical knowledge can be adapted to describe ecological processes and events, you are making the first step into systems analysis.

Above all, if you can't yet program a computer, learn now! The learning isn't entirely painless—it calls for precision of thought, accuracy of expression, a certain degree of pragmatism. Once you have mastered the first steps, you will probably find it much more absorbing and intellectually challenging than cross-word puzzles, bridge, or chess. The facility which you gain will open the doors to much of the 'literature' of systems analysis which will never be published, but which is available in computer programs and algorithms.

References

1. ANSCOMBE, F. J. (1950). Sampling theory of the negative binomial and logarithmic series distributions. *Biometrika*, **37**, 358–82.
2. ANTON, H. (1977). *Elementary Linear Algebra*, 2nd edn. Wiley, New York and London.
3. BAEUMER, K. and DE WIT, C. T. (1968). Competitive interference of plant species in monocultures and in mixed stands. *Neth. J. agric. Sci.*, **16**, 103–22.
4. BALAAM, L. N. (1972). *Fundamentals of Biometry*. Allen & Unwin, London.
5. BEDDINGTON, J. R. (1975). Economic and ecological analysis of red deer harvesting in Scotland. *J. environ. Management*, **3**, 91–103.
6. BLACKITH, R. E. and BLACKITH, R. M. (1969). Variation of shape and of discrete anatomical characters in the morabine grasshoppers. *Aust. J. Zool.*, **17**, 697–718.
7. BRENNAN, R. D., DE WIT, C. T., WILLIAMS, W. A. and QUATTRIN, E. V. (1970). The utility of a digital simulation language for ecological modelling. *Oecologia*, **4**, 113–32.
8. BROSS, I. D. J. (1971). Comment. *J. Am. statist. Ass.*, **66**, 562.
9. BUZAN, A. (1974). *Use your Head*. BBC, London.
10. CHASTON, I. (1971). *Mathematics for Ecologists*. Butterworths, London.
11. CHRISTIE, J. M. (1972). The characterization of the relationships between basic crop parameters in yield table construction. *Proc. 3rd Conf. Adv. Grp. Forest Statisticians, Jouy-en-Josas, IUFRO*, 37–54.
12. COCHRAN, W. G. and COX, G. M. (1957). *Experimental Designs*. Wiley, New York and London.
13. CONVERSE, A. O. (1970). *Optimization*. Holt, Rinehart & Winston, New York.
14. CORMACK, R. M. (1971). A review of classification. *J. R. statist. Soc. (A)*, **134**, 321–67.
15. DAVIES, O. L. (1960). *Design and Analysis of Industrial Experiments*. Hafner, New York.
16. DAVIES, R. G. (1971). *Computer Programming in Quantitative Biology*. Academic Press, London and New York.

17. DE WIT, C. T. and GOUDRIAAN, J. (1974). *Simulation of Ecological Processes*. Centre for Agricultural Publishing and Documentation, Wageningen.
18. EHRENFELD, D. W. (1970). *Biological Conservation*. Holt, Rinehart and Winston, New York.
19. ELLIOTT, J. M. (1973). Some methods for the statistical analysis of samples of benthic invertebrates. *Scient. Publs. Freshwat. biol. Ass., No. 25*.
20. FEDERER, W. T. (1955). *Experimental Design*. MacMillan, New York.
21. FISHER, R. A. (1936). The use of multiple measurements in taxonomic problems. *Ann. Eugen.*, 7, 179–88.
22. FISHER, R. A. (1949). *The Design of Experiments*, 5th edn. Oliver & Boyd, London.
23. FORNSTAD, B. F. (1971). The linear programming planning system of the Swedish Forest Service. *Bull. For. Comm., Lond.*, 44, 124–30.
24. FORRESTER, J. W. (1961). *Industrial Dynamics*. Massachusetts Institute of Technology Press.
25. FOURT, D. F., DONALD, D. G. M., JEFFERS, J. N. R. and BINNS, W. O. (1971). Corsican Pine (*Pinus nigra* var. *maritima* (Ait.) Melville) in southern Britain—a study of growth and site factors. *Forestry*, 44, 189–207.
26. GAUSE, G. F. (1934). *The Struggle for Existence*. Williams and Wilkins, Baltimore.
27. GIBBS, J. N. and HOWELL, R. S. (1972). Dutch elm disease survey, 1971. *Forest Rec. No. 82*.
28. GOODMAN, L. A. (1969). The analysis of population growth when the birth and death rates depend upon several factors. *Biometrics*, 25, 659–81.
29. GOWER, C. and ROSS, G. J. (1969). Minimum spanning trees and single linkage cluster analysis. *Appl. Statist.*, 18, 54–64.
30. GREIG-SMITH, P. (1964). *Quantitative Plant Ecology*, 2nd edn. Butterworths, London.
31. HAMILTON, G. J. and CHRISTIE, J. M. (1974). Construction and application of stand yield models. In *Growth Models for Tree and Stand Simulation*, ed. FRIES, J. Research Notes 30, Institutionen for Skogsproduktion, Stockholm.
32. HARRIS, R. J. (1974). *A Primer of Multivariate Statistics*. Academic Press, London.
33. HILBORN, R. (1975). Optimal exploitation of multiple stocks by a common fishery: a new methodology. *IIASA Research Report RR-75-28*.
34. HILL, M. O. (1973). Reciprocal averaging: an eigenvector method of ordination. *J. Ecol.*, 61, 237–49.
35. HILL, M. O., BUNCE, R. G. H. and SHAW, M. W. (1975). Indicator species analysis, a divisive polythetic method of classification, and its application to a survey of native pinewoods in Scotland. *J. Ecol.*, 63, 597–613.
36. HOLLING, C. S. (1965). The functional response of predation to prey density and its role in mimicry and population. *Mem. ent. Soc. Can.*, 45, 1–60.
37. HUMMEL, F. C. and CHRISTIE, J. M. (1953). Revised yield tables for conifers in Great Britain. *Forest Rec. No. 24*.
38. JEFFERS, J. N. R. (1959). *Experimental Design and Analysis in Forest Research*. Almqvist & Wiksell, Stockholm.
39. JEFFERS, J. N. R., HOWARD, D. M. and HOWARD, P. J. A. (1976). An analysis of litter respiration at different temperatures. *Appl. Statist.*, 25, 139–46.
40. JONES, D. D. (1975). The application of catastrophe theory to ecological systems. *IIASA Research Report RR-75-15*.
41. KENDALL, M. G. (1975). *Multivariate Analysis*. Griffin, London.

42. KENDALL, M. G. and STUART, A. (1976). *The Advanced Theory of Statistics*, Vol. III, 3rd edn. Griffin, London.
43. KLEIJNEN, J. P. C. (1975). *Statistical Techniques in Simulation*. Dekker, New York.
44. KRZANOWSKI, W. J. (1971). The algebraic basis of classical multivariate methods. *Statistician*, **20**, 51–61.
45. LAWS, R. M. (1962). Some effects of whaling on the southern stocks of baleen whales. In *The Exploitation of Natural Animal Populations*, ed. LE CREN, E. D. and HOLDGATE, M. W., 242–59. Blackwells, Oxford.
46. LEFKOVITCH, L. P. (1965). The study of population growth in organisms grouped by stages. *Biometrics*, **21**, 1–18.
47. LEFKOVITCH, L. P. (1966). The effects of adult emigration on populations of *Lasioderma serricorne* (F.) (Coleoptera: Anobiidae). *Oikos*, **15**, 200–10.
48. LEFKOVITCH, L. P. (1967). A theoretical evaluation of population growth after removing individuals from some age groups. *Bull. ent. Res.*, **57**, 437–45.
49. LESLIE, P. H. (1945). On the use of matrices in certain population mathematics. *Biometrika*, **33**, 183–212.
50. LEWIS, E. G. (1942). On the generation and growth of a population. *Sankhya*, **6**, 93–96.
51. LOWE, V. P. W. (1969). Population dynamics of the Red deer (*Cervus elaphus* L.) on Rhum. *J. Anim. Ecol.*, **38**, 425–57.
52. LOWE, V. P. W. and GARDINER, A. S. (1975). Hybridization between Red deer (*Cervus elaphus*) and Sika deer (*Cervus nippon*) with particular reference to stocks in N.W. England. *J. Zool., Lond.*, **177**, 553–66.
53. LUCKINBILL, L. S. (1973). Co-existence in laboratory populations of *Paramecium aurelia* and its predator *Didinium nasutum*. *Ecology*, **54**, 1320–27.
54. MARRIOTT, F. H. C. (1952). Tests of significance in canonical analysis. *Biometrika*, **39**, 58–64.
55. MAY, R. M. (1976). Simple mathematical models with very complex dynamics. *Nature, Lond.*, **261**, 459–67.
56. MAYNARD SMITH, J. (1974). *Models in Ecology*. Cambridge University Press, London and Cambridge.
57. MELLANBY, K. (1976). Mistaken models. *Nature, Lond.*, **259**, 523.
58. NEYMAN, J. (1939). On a new class of "contagious" distributions, applicable in entomology and bacteriology. *Ann. math. Statist.*, **10**, 35–57.
59. OLSSON, R. (1971). A multiperiod linear programming model for studies of the growth problems of the agricultural firm. *Swed. J. agric. Res.*, **1**, 139–78.
60. PEARCE, S. C. (1965). *Biological Statistics: an Introduction*. McGraw-Hill, New York.
61. PIELOU, E. C. (1969). *An Introduction to Mathematical Ecology*. Wiley-Interscience, New York and London.
62. POLLARD, J. H. (1966). On the use of the direct matrix product in analysing certain stochastic population models. *Biometrika*, **53**, 397–415.
63. POLYA, G. (1931). Sur quelques points de la théorie des probabilités. *Annls Inst. Henri Poincaré*, **1**, 117–61.
64. QUENOUILLE, M. H. (1952). *Associated Measurements*. Butterworths, London.
65. RADFORD, P. J. (1972). The simulation language as an aid to ecological

modelling. In *Mathematical Models in Ecology*, ed. JEFFERS, J. N. R., 277–95. Blackwells, Oxford.

66. RORRES, C. and ANTON, H. (1977). *Applications of Linear Algebra*. Wiley, New York and London.

67. RUSSELL, D. G. (1973). *Resource allocation system for agricultural research*. Res. Monograph in Technological Economics 1, University of Stirling, Scotland.

68. SCHATZOFF, M. and TILLMAN, C. C. (1975). Design of experiments in simulator validation. *IBM J. Res. Dev.*, **19**, 252–62.

69. SCHREIDER, G. F. (1968). Optimal forest investment decisions through dynamic programming. *Bull. Sch. For. Yale Univ.* No. 72.

70. SEAL, H. (1964). *Multivariate Statistical Analysis for Biologists*. Methuen, London.

71. SEARLE, S. R. (1966). *Matrix Algebra for the Biological Sciences*. Wiley, New York.

72. SKELLAM, J. G. (1951). Phylogeny as a stochastic process. *Biometrics*, **7**, 121.

73. SMITH, F. E. (1970). Analysis of ecosystems. In *Analysis of Temperate Forest Ecosystems*, ed. REICHLE, D. E., 7–18. Springer, Berlin.

74. SNEATH, P. H. (1957). Computers in taxonomy. *J. gen. Microbiol.*, **17**, 201–26.

75. SPRENT, P. (1969). *Models in Regression and Related Topics*. Methuen, London.

76. STEVEN, H. M. and CARLISLE, A. (1959). *The Native Pinewoods of Scotland*. Oliver and Boyd, Edinburgh.

77. THOMAS, M. (1949). A generalization of Poisson's binomial limit for use in ecology. *Biometrika*, **36**, 18–25.

78. USHER, M. B. (1966). A matrix approach to the management of renewable resources, with special reference to selection forests. *J. appl. Ecol.*, **3**, 355–67.

79. USHER, M. B. (1967/8). A structure for selection forests. *Sylva, Edinb.*, **47**, 6–8.

80. USHER, M. B. (1969a). A matrix model for forest management. *Biometrics*, **25**, 309–15.

81. USHER, M. B. (1969b). A matrix approach to the management of renewable resources, with special reference to selection forests—two extensions. *J. appl. Ecol.*, **6**, 347–48.

82. USHER, M. B. (1972). Developments in the Leslie matrix model. In *Mathematical Models in Ecology*, ed. JEFFERS, J. N. R., 29–60. Blackwells, Oxford.

83. USHER, M. B. (1973). *Biological Conservation and Management*. Chapman and Hall, London.

84. USHER, M. B. (1976). Extensions to models, used in renewable resource management, which incorporate an arbitrary structure. *J. environ. Management*, **4**, 123–40.

85. VAJDA, S. (1960). *An Introduction to Linear Programming and the Theory of Games*. Methuen, London.

86. VAN BUIJTENEN, J. P. and SAITTA, W. W. (1972). Linear programming applied to the economic analysis of forest tree improvement. *J. For.*, **70**, 164–67.

87. VAN DYNE, G. M., FRAYER, W. E. and BLEDSOE, L. J. (1970). Some optimization techniques and problems in the natural resource sciences. In *Studies*

in Optimization **1**, 95–124. Symposium on Optimization, Society for Industrial and Applied Mathematics, Philadelphia, Pennsylvania.
88. VOLTERRA, V. (1926). Variazione e fluttuazini del numero d'individui in specie animali conviventi. *Atti Accad. naz. Lincei Memorie (ser. 6)*, **2**, 31–113.
89. WALOFF, N. (1966). Scotch broom (*Sarothamnus scoparius* (L.) Wimmer) and its insect fauna introduced into the Pacific northwest of America. *J. appl. Ecol.*, **3**, 293–311.
90. WARDLE, P. (1965). Forest management and operational research. A linear programming study. *Management Sci. (USA)*, **11**, B260–70.
91. WATT, K. E. F. (1963). Dynamic programming, 'Look ahead programming', and the strategy of insect pest control. *Can Ent.*, **95**, 525–36.
92. WATT, K. E. F. (1968). *Ecology and Resource Management*. McGraw-Hill, New York.
93. WEATHERBURN, C. E. (1952). *A First Course in Mathematical Statistics*. Cambridge University Press, London and Cambridge.
94. WILLIAMS, C. B. (1964). *Patterns in the Balance of Nature*. Academic Press, London and New York.
95. WILLIAMS, J. D. (1966). *The Compleat Strategyst*. McGraw-Hill, London and New York.
96. WILLIAMSON, M. H. (1959). Some extensions in the use of matrices in population theory. *Bull. math. Biophys.*, **21**, 13–17.
97. WILLIAMSON, M. H. (1967). Introducing students to the concepts of population dynamics. In *The Teaching of Ecology*, ed. LAMBERT, J. M., 169–75. Blackwells, Oxford.

Index

absorption time, 90–1
accuracy, 35, 159, 179
acid rainfall, 160, 161
acidity, of soil, 121
additive, effects, 81–2
  model, 80, 86–7
adequacy, 68, 73, 115
age, of animals, 54–5
  of trees, 41
age class, 50–2, 54–5, 57, 59, 60, 63
age structure, 50–4, 59–60, 141
agencies, 10, 167–8
agriculture, 7, 8, 33–7, 56–7, 96, 145
Agrostis canina, 121
air, 87
alcohol, 28–32
algebra, 16, 22, 49, 50, 65, 90–1
ALGOL, 28, 175–7
algorithm, 26, 28, 115, 138, 145, 163,
  171, 176, 178
altitude, 124
analogue, 1, 49, 60, 67, 95–6, 178
analysis, canonical, 100–1
  chemical, 103
  data, 6, 167, 174
  decision, 163
  discriminant, 101
  multivariate, 100

physical, 103
principal component, 100–10, 118–
  119, 134
  of variance, 77–86
analytic, model, 19–20
  definition, 19
  solution, 3
  technique, 43
angling, 147–8
annual variation, 10, 86
Antarctic, 123
Anthoxanthum odoratum, 121
appeal to nature, 8, 15
applied ecology, 7
applied mathematics, 9, 17, 67
approximation, 27, 43–4, 115, 119,
  178–9
arbitrary function, 39–40, 45–6
Arenicola marina, 107–10, 135–8
arithmetic, 19, 70, 172
Arrhenius equation, 88–9
asymmetry, 71
asymptote, 16
Atlantic period, 56
attribute, 102
autocode, 175
auxiliary variable, 25–6, 29

bacteria, 15–16
Balaenoptera musculus, 54–6
banks, data, 174, 178
  model, 177
bark beetle, 153–4
barley, 33–7, 141
barrage, 103
basal area, 40–1
BASIC, 28, 30–1, 35–6, 40–2, 80–3,
  146, 175–7
beetle, 7, 153–4
Betula, 93
bimodality, 149–50
binomial, 69–71, 74, 76
biological, diversity, 160
  models, 70
  process, 1, 9, 14–15, 48, 88, 96,
    176
  systems, 45
biomass, 14
biomes, 10
birch, 93, 114, 118
birth rates, 57, 70
birthday, 146–7
blocks, 77–86
blue whale, 54–6
bog, 93, 95
Boltzmann factor, 88
botany, 21
bounding, of problem, 2, 3, 5, 13,
  157–165, 168
breeding, 57
Britain, 10, 56, 134, 153
broom, 128

calcium, 103–4, 106–7, 138
calculator, 92, 100, 171–3
calculus, 17–18, 96
calf, 55, 57
calibration, 119
Calluna, vulgaris, 92–3, 95
calorific value, 143
Canada, 153
canonical, analysis, 100–1
  correlation, 110, 133–8, 141
  variate, 123, 128–134

capacity, 20, 178
carbon dioxide, 14, 28
cartography, 120
catastrophe, cusp, 150–4
  fold, 149–50, 153
  theory, 149–155
cattle, 56
causal relationships, 9
cause and effect, 67
centre of gravity, 119
centroid, 125, 127
Ceratocystis ulmi, 153
Cervus elaphus, 56–60
chalk grassland, 86
chemical, analysis, 103
  laws, 96
  processes, 48
  properties, 107, 109–10, 114, 136
  variables, 104, 137
chemistry, of soil, 134
chi-square, 76
Cladonia, 93
classification, 21, 91, 101, 111–12,
  120–2, 127
climate, 7, 86, 158, 160–1
clone, 129
clumping, 69, 71
cluster analysis, 100–1, 110–18
cluster, 71, 111–12, 127
coefficient, 18–19, 27–8, 34, 86, 89,
  102–3, 105–9, 123–5, 127–8, 135,
  142
  definition, 18
colonization, 15, 71
column vector, 49–50
  definition, 49
communication, 21–2, 27, 161, 168,
  176, 178
community, 93
compartment model, 21
compartmentation, 54, 62–4
competition, 7, 33–7, 56, 78
complexity, 3, 6–8, 10, 14, 22, 65, 67,
  90, 159–61, 168–9
component, 24–5, 102–3, 105–10,
  128, 136
composition, 54, 90

computation, 40, 43, 48, 65, 80, 85, 92, 100, 104, 119, 138, 145, 162–4, 171–9
computer, 6, 24–8, 43–5, 48, 60, 65, 80, 92, 100, 115, 119, 142, 146, 163–4, 167, 169, 171–9
languages, 27–31, 34–6, 39–42, 45, 65, 80–3, 146, 167, 169, 171, 175–7, 179
concentration, 160–1
confidence, 86, 166
configuration, 115
conformity, 69
confounding, 160–1
conservation, 8
constant, 15–18, 25–8, 33, 37, 39, 68, 86–9
definition, 18
constraint, 43–4, 56, 65, 78, 84, 90, 138, 142–5, 158, 174, 177
contagion, 69–70
contours, 124
control, 54, 78–80, 146, 176
convergence, 44, 91
Corophium volutator, 107–10, 135, 137–8
correlation, 17–18, 102–5, 107–10, 133–8, 141
Corsican pine, 134
covariance, 86, 124–6, 136
crop, 7–8, 33–7, 41, 141
crowding, 33
CSMP, 27, 29–30, 34–5, 40, 175
CSSL, 27
culture, 28–33, 45, 141
curve-fitting, 40, 42–3, 46
cusp catastrophe, 150–4
cycle, 5, 8–10, 54, 62–4, 159, 166

data, analysis, 6, 167, 174
bank, 174, 178
collection, 10, 92, 167
experimentation, 168
handling, 173
processing, 6, 173
death, 60, 70, 93
deciduous woodland, 10

decision, analysis, 163
making, 1, 4, 6, 24, 61, 146, 167–8
table, 163
decomposer cycle, 9, 10, 159
decomposition, 87–9
deductive logic, 21–2, 48, 65, 162
deer-mouse, 7
deer, 56–60, 128
definition, problem, 2–3, 5, 13, 25, 157–65, 168
systems analysis, 1–10
goals and objectives, 3–4
model, 15
simulation, 15
degrees of freedom, 73–4, 76, 84–5
delay factor, 39
dendrogram, 111, 118
density, 15–20, 34, 36, 59–60, 68, 153
dependence, 91–2
dependent variable, 18–19, 99
Deschampsia flexuosa, 121
description, 12–15, 27, 48, 179
descriptive model, 100–22
design, 19, 77–82, 160, 166
determinant matrix, 131–2
deterministic, model, 9, 15–17, 19, 50, 53, 67, 96–7
solution, 45
deviation, 73, 78, 80, 82–5, 88, 127–8, 160
diagonal matrix, 60
diameter, 40–1, 78–9, 85
dichotomy, 120
differential calculus, 18
differential equation, 15–17, 19, 33, 43, 65
dimension, 102–7, 115–16, 119, 121, 128
discontinuity, 149–50
discrete, 70–1, 89, 102
discriminant, analysis, 101
coefficient, 123–5, 127–8
function, 123–8, 133, 141
discrimination, 123, 125, 127, 129, 131–3, 141
disease, 153–5, 174
dispersion, 69, 71

distance, 69, 71, 111, 114–16, 118–19, 125, 127, 132–3
distribution, 17–20, 68–72, 74–6, 82, 87, 100, 106–7, 109, 116, 127, 153
distribution, 17–20, 68–72, 74–6, 82,
divergence, 149, 151–2
diversity, 160
drainage, 92
drift, 124
driving variable, 43
Drosera rotundifolia, 121
dry matter, 33
Dutch elm disease, 153–5
dynamic, behaviour, 25
    change, 91
    model, 21, 24–46, 48, 65, 67, 96, 141, 175
    processes, 54, 62–3
    programming, 145
    simulation, 138
dynamics, 56–60, 121, 160
DYNAMO, 27, 39–40, 175

earthworm, 158
ecological, process, 2, 4, 7, 91
    sciences, 65
    system, 2, 4, 7, 9, 12–15, 17, 45, 48, 67
economic analysis, 145
ecosystem, 7, 8, 10, 18, 54, 62–4, 158–60, 165
eigenvalue, 50–2, 55–6, 59, 61–2, 64–5, 102, 104–6, 108, 119, 128, 131–2, 136–7, 172
eigenvector, 50–2, 55–6, 61–2, 64–5, 102, 104, 106, 108–9, 128, 132, 136, 172
element, of matrix, 49–51, 53–4, 56, 59–61, 63–4, 128, 132
elm, 153–5
empirical, function, 34–5, 45
    model, 40–3, 96
empiricism, 34–5, 40–6, 96, 112
Enchytraeid, 7
endemic state, 153
energy, 24, 28, 54, 62–4

engineering, 9
England, 154
environmental, factors, 7, 69, 88
    impact, 103, 149
    management, 8
    process, 1
epidemic state, 153
equilibrium, 60, 95–7
ergodic chain, 93
Erica tetralix, 92, 121
Eriophorum vaginatum, 92
error, 28, 84–5, 123, 159, 172, 174
estimate, 18, 21, 30, 44–5, 48, 55, 74–5, 78–80, 122
estuary, 37–40, 107
Euclidean distance, 114
Europe, 175
evapotranspiration, 160
evolutionary operation, 166
experimental design, 19, 79–80, 166
exploration, of models, 1, 45
exponential, form, 88
    growth, 17, 26–7
    model, 16, 53
    notation, 125
    pattern, 16
extractable iron, 113–14, 116–17

F-test, 85
facies, 92–3
factorial, design, 81–2
    experiment, 166
    model, 77, 86
    moment, 20
    structure, 160
families, of models, 4, 9–10, 21, 24, 43–5, 48, 65, 67, 96, 141, 155, 157, 171
fecundity, 50–1, 54–6, 59–60
feed-back, 4, 6–7, 14, 24, 42–4
female, 50–5, 57, 99, 149
    animals, 50–5, 57
fertility, 106–7, 109
fertilizer, 8, 86, 158
fish, 37–40, 146–8
fitting, 18–20, 22, 28, 70
flexibility, 24, 44, 176–7

flour-beetles, 7
flow-chart, 40
flowers, 146–7
fluctuation, 17, 60, 93, 96, 161, 166
fold catastrophe, 149–50, 153
foliage, 33, 68, 87–9, 129–34, 141
forecasting, 43
forestry, 7–8, 40–3, 56–7, 96, 145–6
forest, 40–3, 54, 56, 60–2
formalin, 78–86
formalization, 96
formulation, 5, 10, 17, 48, 164, 175
FORTRAN, 28, 175–7
frequency, 68–9, 71–6, 99, 120, 122, 147
freshwater, 37–9, 71–7
fuel ash, 158
functional, mechanism, 92
    model, 9
    relationship, 27, 42, 87
fungus, 153

Galium saxatile, 121
game theory, 146–9
generating function, 20
generation, models, 13
    solutions, 2, 4, 162
geometric, 20, 71
gestation, 55
goals, 2–4, 13
goodness of fit, 69, 74
gradient, 119
graph, 40, 88, 92, 142–4, 164, 172
grasshopper, 128
grassland, 86
gravity, 119
grazing, 56, 93, 95
    lands, 56
groups, 69–70, 111
growth, 7, 15–17, 26–37, 43, 45, 53, 60–2, 70, 78–86, 93, 141, 145, 160–1

habitat, 7, 9, 56, 71, 96, 120
harvesting, 53–6, 59–62, 141, 145
height, 40–1, 78, 99
Helobdella, 71–7

herbivore, 56–60, 63, 93, 128
hierarchy, 3–4, 13, 111, 121
hinds, 57, 59
homeostatic mechanism, 56
homomorph, 14
Hotelling's T, 125
Hydrobia ulvae, 107–10, 135, 137–8
hypothesis, 6, 8–9, 20, 28–9, 44, 68, 72, 74, 76, 154, 159–60, 162–5, 167–8, 174–5
    of processes, 163
    of relationship, 163
    of relevance, 163
hysteresis, 149–50

identification of goals and objectives, 2–4, 13
identity matrix, 49
immigration, 53, 70
impact, 10, 103, 149, 159–61, 165
increment, 14, 41
independent variable, 99
    definition, 18
indicator species analysis, 119–22
industry, 25, 146, 158, 166
inference, 18, 158
information, exchange, 176, 178
    flow, 25–6
    transfer, 21
insect, 7, 54, 128, 146, 153–4
instability, 8, 52
integral, 26
integration, 28, 167–8
interaction, 6–10, 19, 25, 28–9, 44, 86, 98, 160–1, 163, 165–7, 178
intercorrelation, 104, 108
International Biological Programme, 9–10, 159
interpolation, 35, 40, 43, 45
interrelationship, 4, 7, 13, 22, 25, 134, 137
inversion, 65, 126, 172
invertebrate, 7, 69, 71–7, 103–10, 135–8, 147–8, 153–4, 158
island, 57, 123
iteration, 5, 27, 55, 61, 119, 142, 145

lake, 69, 71–7, 124
Lake District, 113–18
land use, 8–10
larvae, 7
latent root, 61
leaf, 68, 129–34, 87–9, 141
  blade, 129, 132
  litter, 87–9, 141
  shape, 129–34
least squares, 78
leech, 71–7
Lefkovitch equation, 60
Leslie matrix, 55, 62
life, cycle, 54
  table, 57
likelihood function, definition, 19
linear, algebra, 49, 65
  combination, 136
  function, 87–9, 102, 104–6, 123, 141–2
  interpolation, 35, 40, 45
  measurement, 132
  model, 77, 86
  objective function, 142
  programming, 142–5
  relationships, 88, 164
litter, leaf, 87–9, 141
log-normal distributions, 70–1
logarithm, 16, 20, 52, 55, 68, 70–1, 75, 88–9
logic, 14–16, 21–2, 48, 67, 138, 140, 163, 169, 173, 179
logistic, equation, 20
  growth, 26–7
  model, 16
loss on ignition, 103–4, 106, 113–14, 116–17, 138
Lotka-Volterra equation, 33
Luzula multiflora, 121

Macoma balthica, 107–10, 135, 137–8
magnetic, disk, 173–4
  tape, 173–4
male, 54–5, 57, 99, 149
  animals, 54–5, 57
management, plan, 167

manifold, 149–50, 152, 154
manipulation, 15–16, 163, 160
mapping, 14, 67, 107, 109, 120, 123–4, 127
Markov, chain, 89–90, 92
  model, 89–97
material flow, 25–6
mathematical, description, 15, 24
  expression, 16, 26, 140, 163, 169
  formulation, 65, 112
  function, 19
  language, 22
  model, 14–15, 21–2, 24, 43–4, 53, 68–70
  operation, 50
  programming, 142, 146
  relationship, 67
  rules, 14
  technique, 1, 172
  term, 1, 12, 16
matrix, 19, 21, 48–65, 89–91, 93–5, 102–5, 108, 115, 124–6, 128, 130–2, 136–8, 141, 146–8, 172
  algebra, 40–50, 65
  determinant, 131–2
  equation, 50–1
  function, 53
  identity, 49
    definition, 49
  inversion, 65, 172
  Leslie, 55, 62
  models, 21, 48–65, 89, 141
  null, 49
    definition, 49
  multiplication, 172
  square, 49–51
    definition, 49
  symmetric, 49, 132
    definition, 49
  transition, 19
  unit, 49
    definition, 49
maturity, 54
maximization, 140
maximum likelihood estimates, 19, 75
  definition, 19

mean, 68–9, 71, 77, 83–7, 90, 128
mechanism, 92
microcomputer, 65, 178
migration, 37, 40, 55
minicomputer, 65, 178
minimization, 140
minimum spanning tree, 112, 114–17, 133–4
mire, 92–5, 141
missing value, 80, 100–1
model, additive, 80, 86–7
    advantages and disadvantages, 21
    analytic, 19–20
        definition, 19
    banks, 177
    behaviour, 20
    biological, 70
    catastrophe theory, 149–55
    compartment, 21
    complexity, 90, 159–60
    construction, 45, 163–4
    definition, 1, 15
    description, 179
    deterministic, 9, 15–17, 19, 50, 53, 67, 96–7
    development, 159, 168
    differential equation, 16–17
    dynamic, 21, 24–46, 48, 65, 67, 96, 141, 175
    empirical, 40–3, 96
    exploration, 1, 45
    factorial, 77, 86
    families, 9–10, 21, 24, 43–5, 48, 65, 67, 96, 141, 155, 157, 171
    fitting, 18–20, 22
        definition, 18
    formulation, 5, 10, 17, 48
    functional, 9
    game theory, 146–9
    generation, 13
    growth, 15–16, 161
    harvesting, 53
    historical development of, 96
    immigration, 53
    improvement, 1
    linear, 77, 86
    Markov, 89, 97

    mathematical, 14–15, 21–2, 24, 43–4, 53, 68–70
    matrix, 21, 48–65, 89, 141
    multivariate, 19, 21, 99–138
    optimization, 21, 140–55
    parameters, 18
    performance, 167
    predictive, 100, 122–38
    probabilistic, 17
    properties, 17, 95–6
    recursive, 20
    regression, 19, 86–8, 99, 123
    simulation, 19–20
        definition, 19
    stochastic, 16–17, 20, 54, 67–97, 122, 141
    structure, 44–5, 142
    systems, 14
    testing, 1
    validation, 1, 164–6, 92
    verification, 164–6
    world, 12–14, 27, 161–2, 164
moisture, 88, 141, 160
Molinia, 93
monoculture, 28–33, 45
moorland, 10
Morecambe Bay, 103–10, 136–8
mortality, 55, 57
moss, 93
mowing, 86
mud, 106–10, 136–8
mull soil, 87
multiple regression analysis, 86–9, 122
multivariate, analysis, 100
    model, 19, 21, 99–138

Naperian, 68, 75
Narthecium ossifragum, 121
native pinewoods, 120–2
natural, association, 111
    groups, 111
    resources, 145
nature reserves, 8
negative binomial, 70–1, 74, 76
Nephthys hombergii, 107–8, 110, 135, 137–8

Nereis diversicolor, 107–10, 135, 137
Newtonian calculus, 96
niche structure, 160
nitrogen, 103–4, 106, 113–14, 138
noise, 166
non-linearity, 24, 43–4, 87–8, 91, 145, 164
    function, 87
    programming, 145
    responses, 24
non-stationarity, 90
non-zero-sum game, 146, 148
null matrix, definition, 49
nutrient, 9–10, 33, 54, 62–4, 159
nutrition, 114, 118, 143, 147–8

oak, 89
oats, 33–7, 141
objective function, 142–5
objectives, 2–4, 13, 91, 102
observation, 18–19, 111, 123, 128, 138, 149
offspring, 55, 57, 71
operational research, 141
optimization models, 21, 140–55
ordination, 119–22
organic matter, 121, 158
orthogonal(ity), 79–80, 103
    polynomial, 40, 43
oscillation, 20, 53, 65
Oxalis acetosella, 121
oxygen, 14, 87–8, 141

paper-tape, 173–4
Paramecium, 7
parameter, definition, 18
partition, 82–3
pattern, spatial, 67–77
performance of models, 167
Peromyscus leucopus, 7
perturbation, 56, 91–2
pest control, 54, 146
petiole, 129–33
phosphate, 64, 113–14
phosphorus, 62–4, 103, 106, 113–14, 138

photosynthesis, 14, 176
physical, analogue, 1, 96
    analysis, 103
    environment, 103–10, 136–7
    laws, 96
    processes, 48, 67
    properties, 107, 109–10, 136–7
    sciences, 9
    terms, 1, 12
physics, 9, 17, 67, 134, 179
physiography, 134
pH, 113–14, 116–17
Picea sitchensis, 40–3, 78
pinewoods, 120–2
Pinus sylvestris, 93
planning, 149, 167
plot, sample, 78, 80–1, 83–4, 120–2
plotting, graph, 88, 107, 109, 164, 172
Poisson distribution, 68–73
Polya distribution, 70–1
polynomial, 43, 87, 150
poplar, 129–34
population, definition, 18
    dynamics, 149
    parameters, 21
    structure, 52, 56–60
Populus, 129, 133–4
precision, 5, 80, 160, 167, 174
predator, 7, 19–20, 53, 56, 143–6
predictive model, 100, 122–38
presence-absence, 86, 119–20, 124
prey, 7, 19–20, 53, 143–6
principal component, 105, 108, 128, 136
    analysis, 100–10, 118–19, 134
    objectives, 102
principal diagonal, 49, 64, 125, 132
    definition, 49
probabilistic, model, 17
    relationship, 9
probability, of occurrence, 68, 70, 72, 73–4
    theory, 18
problem, bounding, 2, 3, 5, 13, 157–165, 168
    complexity, 3, 6, 161

problem (*cont.*)
  definition, 2, 3, 5, 13, 25, 157–65,
    168
  recognition, 2, 3
  simplification, 3
  solution, 3, 6
process, 2, 9–10, 13–14, 146
production, 33, 61, 141, 146
  forecasting, 43
programming, 27–8, 142–5, 175, 178
program, 28–30, 34–6, 39, 41–2, 65,
    80–3, 87, 146, 175, 177–8
properties, mathematical, 138
  of models, 17, 95–6
pseudo-random numbers, 19
pseudo-spline curves, 43
Pteridium, 93
pulverized fuel ash, 158
punched-card, 173–4

Q_{10}, 89
quadrat, 18, 119–20
quadratic interpolation, 35
quality class, 41

rainfall, 87, 160–1
random(ness), 17, 19, 69–70, 77–8,
    82, 87
  elements, 17
  error, 77, 82, 87
  numbers, 19
rate equations, 25, 39
ratio, 89, 102
reaction, 88–9
reciprocal averaging, 88, 100–1, 118-
    122
recognition, problem, 2–3, 10, 12
recording, 119, 161
recursive model, 20
recycling, 5
red deer, 56–60, 128
regeneration, 60–1
regression, 18–19, 86–9, 99, 122–3,
    172
  analysis, 18, 86–9
  model, 19, 86–8, 99, 123

regressor variable, 18, 99, 122–3
  definition, 18
regressor variate, 123
regularity, 69
relational diagram, 25–7, 29, 37–8, 41
relative space, 33
replication, 80
reproduction, 7, 51, 176
research, organization, 45
  planning, 145
  programme, 167
  strategy, 1, 6, 167
residual, 86–7
resource management, 8, 92
respiration, 14, 28, 87–8, 176
Rhum, 57
river, 37–8, 106–7, 110
rock, 124
root, tree, 33
rotation, forest, 62
row vector, definition, 49

Saccharomyces cerevisiae, 28–32
saddle-point, 146–8
salmon, 37–40
sample, definition, 18
  estimates, 20
  plots, 40, 42
  points, 103, 111
  size, 70
  statistics, 18
sampling, theory, 174
  unit, 68–72, 74–5, 77, 101–2
sand, 106–10, 136–8
sawfly, 7
scalar, 49, 50
scaled vector, 137–8
Schizosaccharomyces, 28–32
science, 9, 14, 22, 178
scientific method, 6, 15
Scolytus, 153
Scotland, 59, 120–2
Scots pine, 60–2
scree, 124
seasonal, 10, 39, 53
seed, 34, 68

seedbed, 78
seedling, 78–86, 93, 146
selection forests, 60
sensitivity, 5, 40, 43, 45, 166–7
    analysis, 166–7
semi-natural woodland, 10
sequential, estimate, 43
    relationship, 13
servo-mechanism, 24
sex, 99
shade, 80
sheep, 56
shell, 106–7, 110
shrub, 87, 120
significance test, 85–7, 103, 106–7,
    109, 132, 137, 166
Signy Island, 123–8
Sika deer, 128
silt, 106, 110
similarity, 111, 118
    index, 118
Simplex, 142
simplicity, 14, 44, 160
simplification, 3, 7
simulation, definition, 15, 19
    model, 19–20
simultaneous equation, 124
single linkage, 111–12, 114, 116
singularity, 149–50, 153
sinuosity, 43
site class, 41
Sitka spruce, 40–3, 78–86
size, 54, 60–2, 69–71, 96
    class, 60–2
    structure, 54, 60–1
social welfare, 145
soil, 8, 78–87, 113–18, 120–1, 134
solution, families of, 4
    generation of, 2, 4, 155, 162
South Georgia, 123
South Orkney Islands, 123
space, 69, 111, 145, 158–9
spatial, distribution, 69–71
    pattern, 67–77
Sphagnum, 92, 121
    Sphagnum palillosum, 121
spruce budworm, 153

square matrix, 49, 51
    definition, 49
stability, 8, 45, 60, 65, 67, 91
stags, 57, 59
stand, of trees, 42–3
standing crop, 14, 42
state, 89–95, 141, 149
    absorbing, 90–1, 93, 95
    closed, 90–1, 93
    successional, 90–1
    transient, 90–1
    variable, 25–7, 29, 37, 39, 41
state-determined system, 26
statistical, design, 166
    distribution, 68
    methods, 77
    technique, 18, 40, 42
    test, 74
    theory, 67
sterilization, 78–86
stochastic, dynamic programming,
    145
    model, 16–17, 20, 54, 67–97, 122,
        141
    process, 20, 97
    relationships, 9, 45
strategy, 1, 6, 143–8, 167–8
stream, 37–9, 69
sub-routine, 176
sub-system, 45, 158–60, 163
succession, 90–5
successive approximation, 44, 52
Succisa pratensis, 121
sugar, 28
sum of squares, 78, 82–5, 87, 128,
    130, 132
sunlight, 33
survey, 6, 18, 22, 77, 103, 107, 120,
    159
survival, 50–1, 53–6, 59, 97
Swedish Forest Service, 145
sycamore, 114, 118
symbol, 14–16, 22, 25–6, 40, 49
symbolic logic, 14–16, 22
symmetric matrix, 49, 132
    definition, 49
synthesis, 10, 159, 167

system, behaviour of, 92
  equations, 27
  operations, 24
systematic bias, 159
systems analysis, definition, 1–10
  phases, 2–5
systems, diagrams, 26
  dynamics, 24–5

taxonomy, 102–11, 128
technology, 178
Tellina tenuis, 107–8, 110, 135, 137–8
temperature, 87–9, 141, 160
territorial behaviour, 69
theory, 9, 44, 67–8, 149–55
thinnings, 42
Thomas distribution, 70–1
threshold, 38–9, 111, 116, 118, 120,
  149
  distance, 111, 116, 118
Tierra del Fuego, 123
timescales, 8, 10, 17
topology, 149
training, 27, 168
transfer rates, 92
transformation, 71, 88–9
transition, 19, 63–4, 89–91, 93–5, 141,
  150
  matrix, 19
transportation, 145
travelling, 143
treatment, 4, 43, 77–86, 158, 160
trophic levels, 54, 62
tundra, 10, 159

unbiased value, 77
uniform, distribution, 68
  spacing, 69
unit, area, 43
  matrix, 49
United Kingdom, 158
universe, 179
upland, 10, 57

Vaccinium myrtillus, 93

validation, 1, 6, 74, 92, 164–7
valuation, 43
variable, definition, 99
variance, 68–70, 74, 77–86, 88, 124–6,
  137
variate, 99–101, 114, 118–19, 122–3,
  128–34, 136, 141
variation, 10, 20, 48, 59–60, 67–8, 86,
  101–4, 107–9, 121, 153, 161,
  166–7
  index, 102
vascular plant, 92–3, 120, 123–8
vector, 49–52, 55–6, 61–5, 94–5, 102,
  104, 106, 108–9, 115, 119, 123–5,
  128, 132, 136–8, 172
  row and column, definition, 49
vegetation, 78, 93, 120, 123–4
verbal description, 12–13, 161
verification, 162, 164–6
Viola riviniana, 121

waste product, 28–9, 158
water, 28, 37–9, 63, 93, 147
  table, 93
weather, 8, 59, 158
weighting, 102, 106
weight, 24
weir, 37–9
whale, 54–6
wildlife, 8
wind speed, 160
woodland, 10, 93, 95
word model, 12–14, 27, 161–2, 164
world, 14, 178–9
  model, 14
worm, 7, 71–7, 153

yeast, 28–32, 141
yield, 33, 40–3
  control, 43
  model, 43
  table, 40–3

zero-sum game, 146